D0070009

COLUMBUS WAS CHINESE

COLUMBUS WAS CHINESE

Discoveries and Inventions of the Far East

HANS BREUER

Translated by Salvator Attanasio

HERDER AND HERDER

WINGATE COLLEGE LIBRARY
WINGATE, N. C.

1972
HERDER AND HERDER NEW YORK
232 Madison Avenue, New York 10016

Original edition: *Kolumbus war Chinese*
© 1970 by Frankfurter Societats-Druckerei (Societats-Verlag),
Frankfurt am Main

ISBN: 665-00001-4
JUL 17 '73
Library of Congress Catalog Card Number: 75-167867
English translation © 1972 by Herder and Herder, Inc.
Manufactured in the United States

Contents

57145

1.
Novae and Pulsars

KAIFENG, 1054 A.D.: silence reigned in the council chamber of the Chancellery for Astronomical and Calendrical Sciences. None of the four scholars present would begin to speak. They were estimable state officials, heirs to important positions, and haste was alien to them. They–the Imperial Astronomer, the Imperial Meteorologist, the Imperial Chronicler, and the Imperial Astrologer–had been brought together by an unusual event in the heavens. Their offices had been in existence for more than a thousand years, and everyone accorded them the highest respect.

Strictly speaking, State positions were not inherited. Their occupants were selected only after they had passed extraordinarily stiff examinations. The examination system was as old as the positions themselves. Before the reforming Minister of State, Wang An-shih, had caught the emperor's ear, the corpus of knowledge required of candidates consisted predominantly of philosophy, poetry, and calligraphy. Knowledge of a practical nature was deemed of slight usefulness only. All that had now been changed: mathematics, jurisprudence, bookkeeping, geography, and agricultural sciences were now obligatory. It now took much longer than ever before for a candidate to prepare for the examinations. Only a few aspirants considered themselves sufficiently prepared before their thirtieth or even their thirty-fifth birthday. In consequence, the power of the cultured and wealthy classes steadily grew; there were few from the lower strata of society who could afford to spend their days in study over a span of so many years.

The examinations were a logical consequence of the teachings of Confucius. All men, with the exception of course of the members of the ruling dynasty, were born equal. Those who acquired the most knowledge should exercise the highest functions in the State apparatus and rule over the less knowledgeable for the sake of the commonweal. Theoretically everybody had the same chance. Yet the poor people, who made up the majority of the population, could not even be aware of them. In practice, the cultured stratum provided all the occupants of high office. Indeed, many families had been doing precisely that for uncounted generations. But each member of the family, as his turn came up, had to pass the stiff examinations in order to be appointed an office.

On the other hand, such was not the case in the Chancellery for Astronomical and Calendrical Sciences. Here the positions were passed on from father to son. The reason for this was that the interpretation of celestial events and the arrangement of the calendar were tasks directly connected with the dynasty. If a new dynasty wanted to monopolize power, it had to control such knowledge. Thus the ruling dynasty made astronomy a secret science. It was not taught at the Imperial Academy, notations were kept under lock and key. Knowledge was handed down from father to son. Outsiders were not allowed even to look into the matter and had no way of knowing that the Chancellery kept records of celestial phenomena going back as far as the year 1500 B.C. These scholars lived in seclusion within the walls of the imperial palace. Here were also located all the instruments they required for their investigations.

Their secluded existence corresponded to the situation prevailing in China at that time. The Sung dynasty was surrounded on all sides by enemies constantly bent upon extending their frontiers at the expense of the Chinese. Consequently, an exchange of ideas or goods with the outer world hardly ever took place. In other words, the Chinese Wall was closed—although not wholly voluntarily on the part of the Chinese. This situation was completely disadvantageous to the peoples living outside this symbolic Wall. Science, art, and technical skills flourished under the Sung. The Renaissance might have been ushered in several centuries earlier had China's highly developed printing technique reached Europe at this time. Medieval knighthood would have fared rather badly in an encounter with the firearms of the Sung dynasty and feudalism might have come to a close even earlier. But the Chinese Wall was closed, and thus the development within the Middle Kingdom remained isolated from the rest of the world. Only the destruction of the Sung dynasty by the Mongols brought down the Chinese Wall.

Let us return to the shores of the Yellow River, to Kaifeng, to the Chancellery for Astronomical and Calendrical Sciences.

Tsi-tan, the astronomer, finally broke the silence: "Messengers have come from Peking, the Observatory has sent news that confirms our observations most exactly."

His hand reached out for the scroll—at that time important messages were still written on silk—and read: "A guest star has appeared in the eastern sky. It appeared in Thiem-Kuan in June, it was as bright as Venus and twinkled in white-red colors. The guest left us after 23 days."

Tsi-tan set the scroll aside. "A rare visitor in the heavens. I have checked the notations on the guest stars and have found the earliest reference in the oracle bones that have come down to us from the Shang kingdom. The text is quite brief: 'On the seventh day of the month a bright new star appeared in company with Antares.' Another bone indicates that the visitor disappeared after two days. That happened 2300 years ago. The archive contains many reports concerning similar celestial events. What is common to most of them is that the guest star sojourned but very briefly and its position in the sky did not

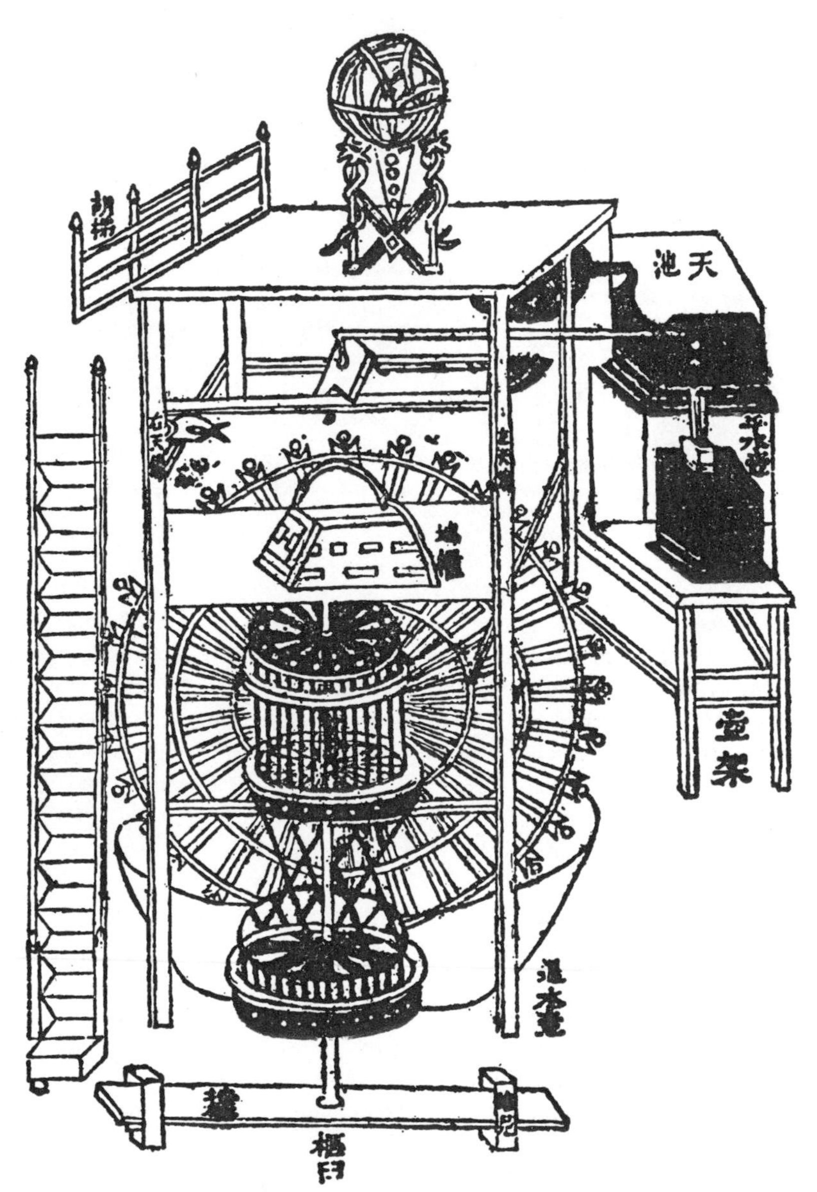

A water-driven celestial globe, published in the works of Hsin I Hsiang Fa Yao.

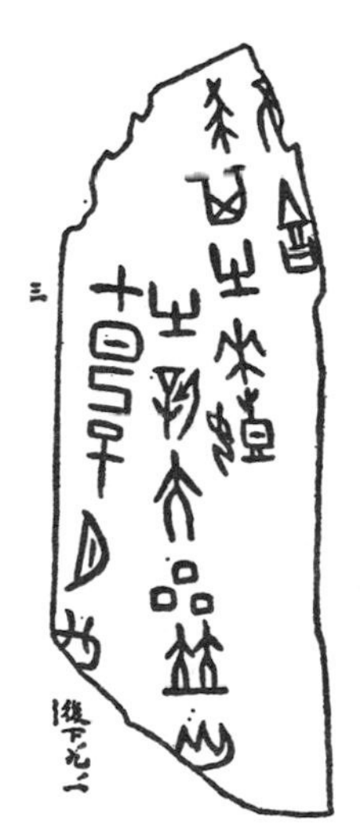

Diagram of an oracle bone from the fourteenth century B.C. (Shang Dynasty). This is the oldest reference to the appearance of a nova in the constellation Scorpio.

change. It always remained in the same lunar mansion. What, then, is the significance of the guest star?"

Yang Wei-te, the Imperial Astrologer, now spoke up: "The guest star in Thiem-Kuan for a time emitted yellow rays, hence it is of significance for the Emperor. Aldebaran was not infringed upon. My calculations show the Emperor's power lasting for a long time still and the country's wealth increasing."

The scholars now bent over the celestial charts. A twentieth-century observer would not have recognized the constellations. In China the individual heavenly bodies were combined with quite different symbols from those of the Greeks and Arabs, whose constellations we have to a great extent made ours. Only the Great Bear appears to us in the familiar form. The charts contained 283 constellations all together and the exact positions of more than 1500 individual stars.

In January of 1054 astronomers in China used the charts of

Chen Cho, dating from the fourth century. Chen Cho had collated them on the basis of the star catalogues of the astronomers who lived in 400 B.C.–Shih Shen, Wu Hsien, and Kan Te.

Presumably there were even older star catalogues, but these have not been preserved. The oldest Chinese description of a star still known to us is to be found on an oracle bone from the fourteenth century B.C.

The coordinates system was amazingly modern. The positions of the celestial bodies were measured with reference to the equator, whereas the stellar locations on the oldest western catalogue, namely, that of Hipparchus (c. 190-125 B.C.) were measured on the apparent solar orbit, on the ecliptic system.

This essential difference in the method of recording heavenly

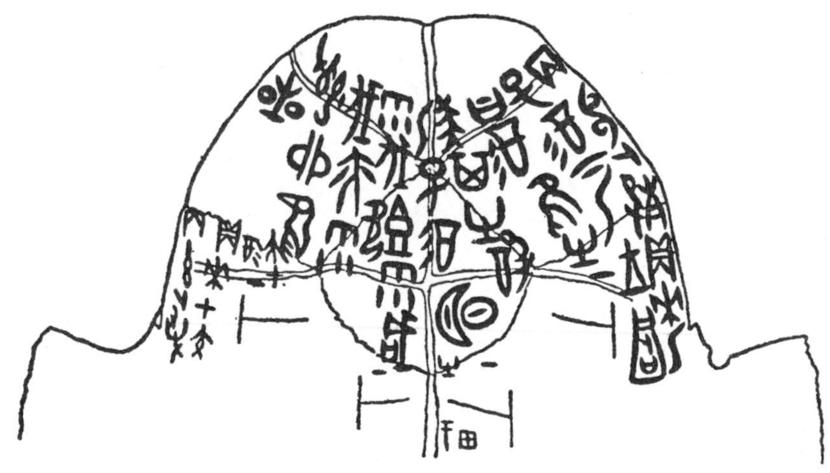

This oracle bone bears the oldest description of an individual star. It is Alphard ("The Isolated One") in the constellation of the southern Hydrae. The inscription is about 3300 years old.

bodies on a chart supports the thesis that no knowledge of classical Greek astronomy ever reached as far as China. Both culture areas, however, presumably derived impulses from the older Babylonian astronomy.

A few years after the Council of Kaifeng, the chronicler Su Sung was commissioned to draw new astronomical charts. After six years of labor he completed five charts. Two of them, delineated cylindrically, looked very much like those of Mercator. They are still quite noteworthy, because, presumably, they were the first printed astronomical charts. Thus nothing stood in the way of their diffusion on a broad scale.

We also find celestial globes and circumpolar astronomical charts already in use at this time. These are charts in which

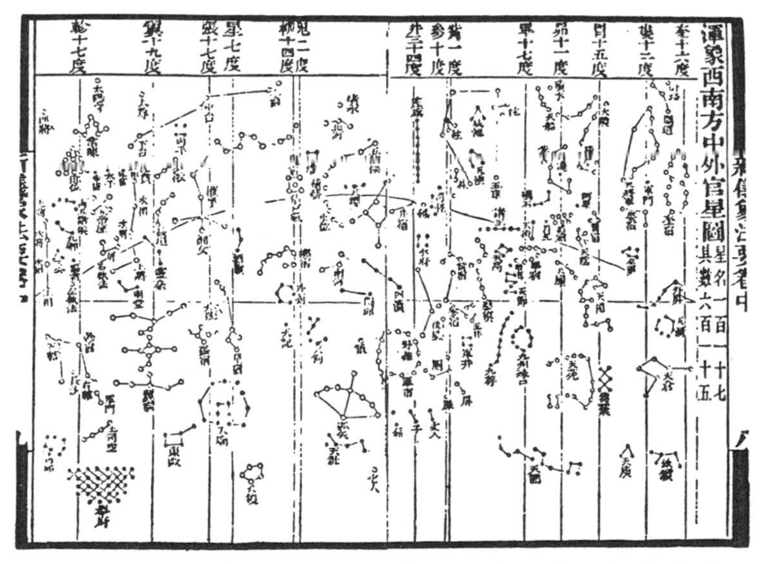

Chinese astronomical chart, 1092 A.D. The vertical crosslines divide the lunar mansions.

stars, or other heavenly bodies, are represented around the pole star as center. An especially beautiful and accurate exemplar is the planisphere that was drawn for the education of the Prince who ascended the throne in 1195 as Ning Tsung and who ruled the kingdom of the southern Sung dynasty for 29 years.

Dallas, 1968: The auditorium–Section B, Pulsars–of the Symposium on Relativistic Astrophysics is packed from wall to wall. The avid audience had come to hear a lecture on the fascinating property possessed by some radio stars which were emitting extremely brief and regular impulses. Rumors were circulating that astronomers at Flagstaff Observatory in Arizona had made observations that would shed an utterly new light on pulsars.

Silence fell over the lecture hall when the speaker took his position in front of the projection screen, stuck the microphone in the breast pocket of his sport shirt, and began. We shall summarize his talk for the reader:

In 1946, radio astronomers–still a numerically small group–discovered the first extraterrestrial radio sources. This discovery was not made through light-susceptible telescopes. Instead, radio astronomers use gigantic antennae for their investigation of celestial phenomena. A point in the constellation Cygnus was emitting radio waves. In the course of the year the number of individual radio sources (sometimes also called radio stars) rose to several thousands. Now if the position of a source of such a kind is investigated with an optical telescope, the observer occasionally discovers a very faint star in the very same place.

After overcoming their initial surprise, astrophysicists ceased to consider the existence of astronomical radio sources as unusual phenomena. They were, in fact, celestial objects that radiated their energy in a longer wave-length region than obtains with known stars. Soon, however, particularities were discerned: some of these sources seemed to be located outside our Milky Way system, so that their energy production had to be a millionfold greater than that of the sun. These celestial objects called quasars (quasi stellar objects) saddled theorists with new riddles. The astrophysicists were really flabbergasted, however, when in February of 1968 it was observed that one of these radio sources did not emit a temporally uniform radiation. On the contrary, it flashed with an extraordinary precision. Through a radio telescope we observe pulsars exactly as a sailor on the horizon sees the beam of the lighthouse in equal time intervals. A radio impulse appeared for the duration of some milliseconds, followed by stillness. After a few seconds there was a new signal. The interval between the individual bursts is extraordinarily constant. For example, with the radio source CP 1919 it is quoted as 1.33730113 with an uncertainty of 7 in the last figure.

Up to now thirty-six such pulsars have been found, and each one possesses its own rhythm.

"And gentlemen," said Professor Grey, concluding his lecture, "one of these pulsars is the radio source NP0532, Baade's Star in the center of the Crab Nebula."

There it was again, the Crab Nebula in the constellation of Taurus. If we investigate the red shift of the individual regions of this amoeboid gas nebula, a big surprise is in store for us: gas particles, proceeding from a common center, move at a velocity of several thousand kilometers per second. This center is occupied by a faint star named after the astronomer Walter Baade. From all sides the nebula rushes outward from this center like the gas clouds of an explosion. On the basis of the measured velocity and the distance of the gas cloud from its origin, it is easy to calculate when the explosion occurred and pinpoint it in time: 1054. If we consult the old Chinese notations from Kaifeng and Peking, to our utter astonishment we discover that the guest star had appeared at this time and in the same position in the sky. This means that the event observed and recorded in China, and in Japan as well, was a supernova, a gigantic stellar explosion. Today we discern the debris of this cosmic catastrophe in the Crab Nebula.

Reliable reports concerning this event came only from the

Far East. That is not surprising. At this time the exact sciences in the West did not rate very high. Their practitioners were not asked about it even though the first scientific writings of classical antiquity had already been rediscovered, translated, and copied. Alchemy alone could hope for official encouragement and tolerance. After all, the European dynasties were constantly in financial straits; a formula with whose help mercury could be transmitted into silver or even into gold would have indeed been royally welcomed. Not to mention, of course, the elixir of life that is in special demand when war is the order of the day. Indeed, there was no lack of fighting; Henry III had widely expanded the borders of the German Empire and now it was a question of defending them.

If a knight was not engaged in an ongoing military campaign, he had to risk his life in a tournament anyway. Hence an alchemist was a welcome personage. He would create gold for armaments, weapons, finery, spices, and the construction of fortifications. Moreover, with his immortality drug he would preserve the barons from all harm to life and limb. In those days only a truth-seeking scientist had a hard time earning his daily bread.

Arab sources are also silent about the nova of 1054. Perhaps the corresponding notations have not yet been discovered, or they have been lost forever. Arab astronomy, in particular, was highly developed at this time, and star gazers in the many observatories between Granada and Samarkand kept a vigilant eye on celestial phenomena. So striking an event as a new star which suddenly disappears after a brief time could have remained hidden from the Arab star gazers only with great difficulty.

Is it possible that the supernova was also observed by the Indians of North America? Recently rock drawings were found in a canyon in Arizona which can be interpreted in this sense. The crescent moon is depicted over a round object. The latter cannot be the sun since it would completely eclipse the new moon. Is it a likeness of the supernova? Archeological investigations prove that Indians lived in cave dwellings of the canyon between 900 and 1100 A.D. Further, it has been calculated that the moon in July of 1054 actually showed a crescent and, therefore, was located close over the guest star in the sky. All indications suggest that the Indians actually resorted to this unusual means to depict a celestial event.

On the other hand, the two other known supernovae were also observed in Europe: in 1572 by Tycho Brahe in Cassiopeia, the constellation in the northern hemisphere, and in 1604 by Kepler in the Ophiuchus, the constellation astride the equator. Both new stars are likewise recorded in the annals of China, Japan, and Korea.

Supernovas are exceedingly rare, whereas novae, that is to say, stellar explosions of a lesser intensity, appear more frequently in the sky. In most cases these new stars are visible for only a few weeks or even for but one day. Then their brightness dims and they can no longer be seen by the naked eye. In most cases a radio wave appears in the same place.

Meanwhile, back in Dallas the discussion was in full swing: "If we relate the pulsars to the novae, their reciprocal correlation could be established with certainty. Is this the case?"

"Almost all particulars on the novae of the last 3000 years came from the Chinese sources, if we disregard the time from 1600 onward," replied the speaker, while leafing through his notes at the same time. "Lundmark has been searching for the corresponding data in the catalogue of the historian Ma Tuan-lin. The spatial distribution of the guest stars noted by him accords with the average distribution of the novae observed in modern times. Hence we have no warrant to doubt the reliability of these notations. As a further example of the precision of Chinese chroniclers, I should like to point out that their lists of solar and lunar eclipses up to now have withstood every critical testing. Our colleagues of that time were altogether careful, and we can build on their findings."

Let us leave this lively discussion and return to Kaifeng, the capital of the Sung dynasty.

The assembled scholars were a part of the hierarchy, highly paid State officials who enjoyed the privilege of inheriting their positions. Let us observe them more closely and ask why they were so important. The great importance of the Imperial Astrologer is clear enough. He read the future from the constellations of heavenly bodies. He interpreted what the stars promised for undertakings that were in the planning stage, and gave advice on lucky and unlucky days. In short, he performed those services for which the incorrigible in our day still pay their astrologers. But what about the other members of the Chancellery?

They were primarily responsible for the making of the calendar. The calendar, which had to be approved each year by the emperor, played an important role in an agrarian country like China. It listed the beginning of the monsoon, of such vital importance to the peasants, and fixed the time of the melting of snows, thereby forewarning of spring floods that might be expected. It contained particulars concerning bad and good days in the year and recommended ways of warding off evil spirits from the hearth. The oldest Chinese calendar that has been preserved is now 2500 years old. Besides purely calendrical particulars, it also contains hints on animal breeding and edifying prose pieces, just like a modern farmer's almanac.

In the Sung period calendars, of course, were printed. The characters were incised into smoothed wooden boards from which paper copies were drawn. A simple process and, if one got his hands on the blocks, one could conduct a lucrative trade in calendars. This was particularly true for those who brought the new calendar among the people at the earliest possible opportunity. The inhabitants of the province of Szechuan already were reputed to be an enterprising folk, so the idea occurred to some of them to sell the calendar before the emperor had officially approved it. This, of course, was illegal and strictly prosecuted by the police. Nevertheless, under-the-counter sales were brisk. This is attested by the corresponding report on this illegal traffic forwarded to the Imperial Court by local administrative officials.

The astronomers watched over the chronological course of the year and attentively followed events that transpired in the sky. Solar and lunar eclipses were especially spectacular. Here the written traditions reach far back in time: six lunar eclipses (1361, 1342, 1328, 1311, and 1217 B.C.) and a solar eclipse in 1217 B.C. are listed in the oldest extant records. These lists were preserved almost unbrokenly from the eighth century B.C. up to the Middle Ages.

Here the legitimate question rises as to whether at least a part of these records may not have been falsified. That would not have been unusual since the official chronicles were often "rewritten." By so doing one could most conveniently praise the governorship of the ruling dynasty of the moment and blacken the reputation of its predecessor. (Indeed this is a practice that is still vigorously in vogue all over the world.) It is difficult to uncover falsifications of this kind, especially when several centuries have transpired in the meanwhile. In scientific records, however, the danger of falsification is extremely lessened. No doubt astronomers and calendar makers were persons of importance, but there was no social prestige to be gained with sciences that were in the main oriented to the practical affairs of life. In Chinese society it was a matter of indifference whether an astronomical discovery, a scientific cognition, or a technical invention was made today or thousands of years ago. Since neither prestige for a dynasty nor for a person could be gained through forgeries of this kind, it can be safely assumed that these records have probably not been tampered with.

Astronomers find it more satisfying to predict events rather than merely to record them. That is asking too much from an age in which the moon's orbit in the sky could not be calculated. Nevertheless, eclipses recur in fixed intervals which we can find in old reports. The so-called Saros Cycle was known in Babylonia as early as around 500 B.C. It predicts that an observed eclipse recurs after 223 synodical months (about 18 years and 11 days).

The Greeks also used this method, taken over from Babylonia, to predict occultations. Although it has been established that Chinese astronomy took over several features from the

WINGATE COLLEGE LIBRARY
WINGATE, N. C.

Babylonians, the Saros Cycle was never employed in China. Another method was applied which was also quite accurate, based on a period of 135 months. In the second century solar and lunar eclipses could already be predicted exactly to the day, later to the very hour, which was of extraordinary importance to a rustic population engrossed in mystical representations of the world. If the eclipse was indicated in the calendar, evil spirits could completely swallow the sun and rob men of light. Only a numerically tiny stratum was educated enough not to see signs portentous of disaster in eclipses. The peasants in the majority adhered to Taoism. This religion had developed from old shaman cults and early knowledge of nature, and attributed great importance to all natural phenomena. Its adherents, understandably, reacted less rationally to unusual celestial events. At times, of course, the moon or the sun darkened without any prior proclamation by the astronomers, producing a great panic among the populace.

We read over and over again that, even before the adoption in Europe of the chronology of the Christian era, a Chinese emperor ordered the beheading of his court astronomers because the capital was taken by surprise by an eclipse. But this is a cock-and-bull story; the Chinese emperors, or at least their advisors, were educated enough not to demand the impossible. An amazing event that took place in the year 145 B.C. is authenticated: an expedition set out from the capital, bound for Shantung, to observe a predicted solar eclipse which was visible only from the highest points of this peninsula. Today astronomers pursue by plane the lunar shadows extending over the earth in order artificially to prolong the occultation and thereby gain more time for observation.

There was still another viewpoint which explained why eclipses aroused such a great interest among astronomers: they appeared only with an exact full or new moon. Thus the beginning or the middle of a month could be precisely determined. The accuracy was very important for calendar making. After all the main job of the Imperial Observatories—there were two in the capital so that observations could always be compared—was the arrangement of the next calendar. This was a commission of great importance inasmuch as the person who deemed whether or not the calendar was officially correct was considered to be the emperor of the country. Everyone who accepted the regulation of the calendar, by definition, belonged to the Middle Kingdom; he was not a barbarian but instead lived within the compass of the Chinese Wall.

When a new dynasty came to power one of its first official acts was to declare the old calendar null and void. A new calendar with new characters immediately appeared. Private scholars (who abounded in old China) took care not to engage upon calendrical calculations because one could all too easily be suspected of conspiring with a future usurper. Thus astronomy, so closely bound up with the calendar and astrology, remained more or less an arcane science pursued by reliable servants of the State.

Meanwhile in Houston, Professor Thomas Gold of Cornell University advanced his pulsar theory. It had begun forty years before with the theory of neutron stars postulated by Walter Baade and Fritz Zwicky. This was defined by them as the end phase of individual stellar development. At a very late stage, namely that of the "White Dwarfs," a star suddenly compresses into a neutron star, with a diameter of no more than a dozen kilometers, with an enormous release of energy. The matter in the star's interior assertedly is of a fantastic density, a cubic centimeter weighing several million tons.

"Stars of this kind cannot be observed with optical instruments, since they emit no light in the visible spectrum. But they come under consideration as sources for radio waves. The diameter, unusually small for celestial objects, warrants an unforced explanation of pulsars as spinning neutron stars," continued the scientist. He pressed the remote control of the projector.

"Here you see schematically a spinning neutron star surrounded by its magnetic field. Plasma breaks out of the

The Hsi and Ho brothers are commissioned by the Emperor Yo to prepare a new calendar.

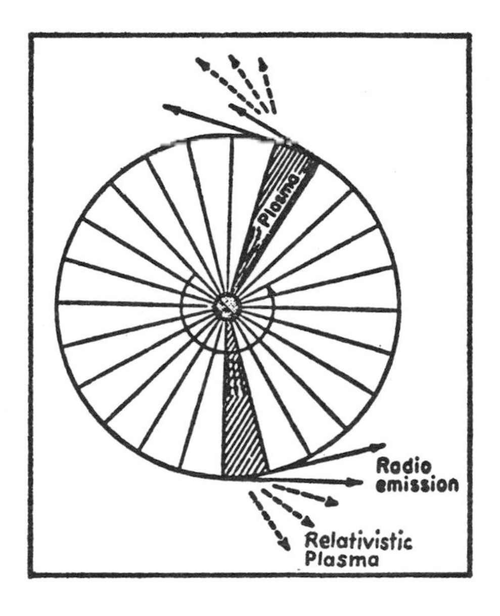

interior, arrives in the magnetic field rotating at the highest velocity, is accelerated there at relative velocities, and expelled. We observe these periodic bursts as pulsars. A concrete analogue would be the rotating searchlight of a lighthouse which blinks in regular intervals for the stationary observer. Let us turn now to the mathematical support of my assertions."

We shall spare the reader this passage and, instead, recapitulate what the Chinese records say about some variable phenomena in the sky.

In addition to guest stars, which do not change their position, comets–those visitors from outer space–also aroused a special interest. Comets, after all, were always considered as presages of important events. The time of their apparition, the size of the tail, the color and orbit were exactly determined. So precise, for example, were the data concerning a comet from the year 11 B.C. that in the modern age astronomers could calculate the orbital elements from the Chinese records. This was Halley's comet, a member of our solar system that comes into range of sight every seventy-six years. There can be no doubt that it was the same comet described in the Chinese chronicles of 240 B.C.

In the West comets were always considered presages of impending calamity. According to the Ptolemaic system in which the crystalline celestial spheres were viewed as perfect, the visitor from space was bound to introduce disharmony into the cosmos and, as a result, also throw the human race out of joint. Sunspots also did not accord with that symbol of purity, the radiant sun. Anyone who claimed to see such things had to be suffering from delusions; what ought not be simply cannot be. If a sunspot was so distinct that everybody could see it through a piece of lamp-blacked glass, it could be only a question of an orbital transit. Truly, the scientific mode of observation and thinking made its way into the medieval West only lumberingly and laboriously.

The situation was quite different in China. Here no State religion stood in the scientists' way. The Establishment honored Confucius less as the founder of a religion than as the author of a teaching for the proper guidance of the State and of men. This teaching did not promote the sciences but neither did it stand in their way. Thus since 28 B.C. the Imperial Chancelleries kept registers of observed sunspots that were value-free notations.

The cosmological world view "Hsuan-Yeh," originating as early as the third century, could hardly be more modern: sun, moon, and stars float through the otherwise empty sky exactly like comets without reciprocal relations. A guest star is simply one that accidentally has fallen in the proximity of the earth, from which it will again draw away. Hence why should its coming always bring disaster? The exclusive concern of the star gazers was to pinpoint the celestial visitor's position. The time periods with which Chinese astronomers dealt also strike us as very modern. They did not find it incomprehensible that celestial events could lie in a past going back a hundred million years.

Professor Gold neared the end of his lecture: "Why pulsars were not observed earlier is clear: the signals received were so weak that up to now they were lost in space as individual events. The impulses first had to be integrated over long periods of time if they were to be forced out of the acoustic background. Only recently have deep-freeze molecular magnifiers and larger antennae with better post-control systems made it possible to investigate the frequency of the impulses in the region of seconds and, in the meantime, also for microseconds. I can say that the discovery of pulsars is one of the most important and exciting events of our century."

The lecture closed to enthusiastic and continuous applause. Radio astronomers of scientific institutes in America, Europe, and Australia exchanged experiences and presented their observations and interpretations for discussion. Only the representatives of the gigantic laboratories of the American Air Force were silent. They were working for the Government and many of their findings were "classified" as the matter was put. A potential enemy after all could draw conclusions about the quality and accuracy of the installed electronic devices even from seemingly so innocent radio astronomy data. The word "secret" was stamped on the envelope of their records. Only the Government decided what information could be published as not crucial to national defense. In this respect Pentagon officials are not known for their generosity; the decree from the year 840 A.D. mentioned earlier in this chapter would doubtlessly be approved by them.

The secrecy surrounding astronomical records led to misunderstandings with the European visitors who showed up in China as early as the thirteenth century and, ultimately, to the decline of Chinese astronomy. Astronomy was not taught; obviously there were no records, *ergo* the Chinese were wholly ignorant in the field–this was apparently the inescapable conclusion that was drawn. In addition, the Chinese Wall was hermetically sealed under the Ming dynasty (1368-1644).

Visitors from other countries were not welcome, for it was believed that nothing good or useful could come from their presence. The Italian Francesco Carletti who conducted trade along the Chinese coasts at the close of the sixteenth century, described this attitude as follows:

"They believe that they have a superfluity and an excellence of all knowledge so that they really are needful of nothing. Therefore they have legally forbidden–indeed by invoking the death penalty–that anyone leave the country or enter the country. The only persons exempt from this law are those who come as envoys of a neighboring tributary. Another exception is constituted by those who declare that they are bringing a tribute of some kind or any other gift, as do many merchants who travel on the land route from East to West up to India. Others come under the pretext of being philosophers and of being desirous to learn something among them." (From "Journey Around the World," 1594.)

This time it was not the "barbarians" but the Chinese themselves who were the victims of the closed Chinese Wall. While within China astronomy atrophied in the traditional representations, European scientists such as Galilei, Tycho Brahe, Copernicus, and Kepler had transformed astronomy from an empirical science into an exact science. On the basis of the quantitative laws they derived, eclipses now could be calculated with an accuracy hitherto unknown. Calendar makers profited considerably from this knowledge. But China would hear nothing of all this; she dropped the thread of contact and for several centuries was no longer to find it in astronomy.

In the middle of the seventeenth century, when the first Jesuit missionaries arrived in Peking, the capital, their astronomical and calendrical knowledge was far superior to that of the Chinese. These discrepancies were further underlined by the fact that many records were kept from the Jesuits since astronomy was still considered as a science the knowledge of which had to remain within the walls of the Imperial Palace. In 1669 the missionaries proposed a reform of the calendar. The weak ruler lent them a willing ear and the Western calendar was introduced. Chinese astronomy was not to recover again from this break with a 3000-year-old tradition until the twentieth century.

The sensation of the astronomical congress in Dallas was the lecture of a member of the Steward Observatory in Arizona. The Baade stars, the remnant of the supernova of 1504, had been more accurately investigated there by a light-susceptible telescope. In the course of the investigation it turned out that it flashed in the same rhythm as pulsar NP0532; both radiation sources , therefore, had to be identical.

Up to then the light of the star in the center of the Crab Nebula, flashing in second intervals, had escaped attention since the photographic plates were always exposed to solar rays for too long. It was the synchronization with the corresponding radio radiation that first showed that this pulsar also flashed in visible light. And, according to Thomas Gold's theory, this was not possible. The optical pulsar will certainly stimulate further theoretical investigations.

What a welcome opportunity for the writers of science fiction, who can portray the pulsars as cosmic lighthouses of superior civilizations whom they serve as navigational aids in journeys through the cosmos. One thing in the meanwhile, no doubt, is generally known: man is no longer the only intelligent being in the universe. It would be presumptuous to attribute to man an exceptional position. Is this a cognition that has first been acquired in the age of space travel? By no means. Seven hundred years ago in Peking, Teng Mu speculated:

"Heaven and earth are large, yet in the whole cosmos they are no more than a tiny grain of rice: it is as if the cosmos was a tree and heaven and earth one of its fruits. The tree is like a kingdom and heaven and earth no more than an individual person in this kingdom. A tree bears many fruits and one kingdom embraces many people. How unreasonable it would be to assume that there were no other heavens and no other earths beside the heaven and earth visible to us!"

The high officials of the Imperial Chancellery for Astronomical and Calendrical Sciences would certainly have been highly gratified had they suspected the importance that would be attributed to their observation nine hundred years later. After the reports on the guest star had safely been stored in the repository, they discussed their next task. The astronomical charts lying on the book shelves in the meanwhile had no doubt become too old and, moreover, were drawn by hand. They composed a petition to the Emperor requesting him to authorize the drawing and printing of new astronomical charts. They also mentioned the plan of having a huge astronomical clock built on the palace grounds. This, of course, would require extensive preparations. The Tantric Buddhist monk I-Hsing and the horological engineer Liang Ling-Tsan had left behind exact descriptions of their clock rotated by water power. Meanwhile, the old scroll had been lying in the archive for the past 329 years. Nevertheless, the additional calculations would require much time.

2.

Abacus, Binary Numbers and Magic Squares

MAJOR Cunningham rubbed his eyes in utter disbelief. Could his sergeant with the electrical calculating machine be slower than this gossamer-thin Japanese bank employee? All he had for making his calculations, after all, was this curious wooden instrument. What the devil was it called, anyway? Yes, of course, the soroban. By this competition the major, a true-blue Connecticut Yankee, wanted to show these Japanese that superiority did not lie only with American weaponry. It was the year 1946 and "re-education" was the order of the day, so there could be no harm in also demonstrating the pre-eminence of American commercial technology.

The competition, which drew scant public notice, took place in Tokyo: modern calculating machines against the ancient, traditional abacus. Major Cunningham called out 6-digit numbers, and then named a simple mathematical operation. The fundamental arithmetical operations, addition, subtraction, multiplication, and division, alternated in lively succession. The fingers of the sergeant—who was the assistant paymaster of the regiment—pressed the keys of the calculating machine, the register clattered, and intermediate results appeared on the calculator. Under the thumb and index finger of the Japanese, the lenticular counters of the abacus flew back and forth and clicked against the middle section of the soroban. Almost always, the nimble fingers arrived at the left end of the abacus before the electrical machine came to a stop. Victory over modern technology?

This competition and its outcome are not fictitious The traditional abacus (called *soroban* in Japan) proved inferior to the electrical desk calculator only in the multiplication operation. This disadvantage was more than balanced by the greater accuracy of the abacus results. Today, a quarter of a century later, the result of a similar competition would be quite different. The Japanese bank employee would not stand a ghost of a chance. In the meantime, modern electronic consecutive switching devices have replaced the mechanical indicators of the old desk calculator. Nowadays mathematical operations with ten-digit figures require speeds that are measured in milliseconds. Nevertheless, travelers to Japan and China still see this simple device in banks, commercial houses, currency exchange offices, in the offices of steamship lines, and in restaurants. It also still plays a role in Russia as a computing aid, where it is called *stschoty*. Made out of wood and bamboo, it is cheap, and in the hands of experienced persons it is as good as the desk calculators which are a thousandfold more expensive. The greatest disadvantage of the abacus, however, is that the intermediate results are lost immediately.

Actually, the abacus is only a refined form of the device on which the reader may have learned the rudiments of addition and subtraction in elementary school. Whereas in the West ten beads are threaded on a wire and can be moved left and right, in the Asiatic variant only five tiny moveable discs are threaded on a wire. From twelve to thirty wires are so arranged as to form a series of parallel columns. This is the form most used today; in the past every conceivable variation existed, from the 4-plus-1 system to twenty pebbles on one row.

Chinese sources seem to refer to the use of the abacus as

early as 220 A.D. The word abacus stems from the Arabic, an indication of how these calculating machines could have reached Europe. It is first mentioned there in the eleventh century by Hermann the Lame. It is, however, just as possible that the abacus was invented in the West, independently of other sources. Obviously a simple device of this kind could have been developed by any civilized human grouping. Merchant travelers like Marco Polo certainly made the acquaintance of these computing aids on their journeys, and perhaps even brought them back with them. This would explain the variety of the forms of the abacus that appeared in Europe.

There were computing aids of more simple design in China and elsewhere a long time before that. One of them is even innate to man: the ten fingers of his hand. If we reckon on our finger joints instead of on our finger roots we can count past 10. Long and short sticks, rods, which in each instance represent an arithmetical unit, appear as the logical next step.

There are many old Chinese writings that make mention of the deftness with which this or that person operated counting rods: "They flew so swiftly back and forth across the counting board that they utterly confounded the eye." The center of the Chinese character for "calculation" seems to depict a board of this kind on which the sticks were manipulated. Later numerical figures for making calculations and for bookkeeping purposes came into use which no longer resembled the rod numerals. From the sixteenth century on Chinese merchants used their own forms so that their figures could be represented in at least three different ways—still other characters appeared on coins.

The counting rods are found once more in the hexagrams of the Book of Changes (I Ching). The Book of Changes is a curious mixture of old peasant rules, magic formulas, maxims of practical wisdom, aphorisms, incomprehensible mysticism, in short: a competitor of the astrological book of charms of the Middle Ages or of modern times for that matter. Generations of soothsayers have lived from their efforts to interpret its obscure statements, present them in a comprehensible form, and explain everything from them. We recognize, therefore, that the Book of Changes, whose oldest material was presumably written in the third century B.C., in ancient China served the same purpose as, for example, the sixth book of Moses does today.

We can go even deeper into this speculation and find ourselves in excellent company: Wilhelm Leibniz (1646-1716), philosopher and mathematician has preceded us. The Jesuit priest Joachim Bouvet brought a copy of the Book of Changes with him out of China. Leibniz corresponded at length with the missionary concerning it. Only a few years earlier the Leipzig savant had developed the binary notation system. Now to his amazement he believed that he had found it in the 2000-year-old work.

This interpretation of the illustrious savant caused a sensation in European scientific circles. Countless theories and attempts at interpretation were undertaken. If binary arithmetic was already known in ancient China, what discoveries might they have made even before that? The excitement soon died down, however, since the trace found in the Book of Changes led nowhere. No other passage in Chinese scientific literature bore up the hypothesis that the Chinese had knowledge of the binary system.

What about present-day interpretation? Presumably the creator of the hexagrams did not have binary numerals in view. Rather, his symbols are exclusively the possible combinations of two different characters within a definite sextuple arrangement.

The hexagrams in the Book of Changes sank into insignificance, but not Leibniz's dyadic arithmetic. Indeed, after the Bad Hersfeld engineer Zuse in the 1940's developed the electronic data processing machine the computer, the binary system proved its great practical value in the presentation of figures. Many electronic elements, whether relays, or ferrites, exhibit two states: they are switched on or off. If the designations 1 and 0 are coordinated with these states, the figures can be represented, fed, and manipulated in binary form through a great number of these electronic elements. Modern computers contain millions of elements of this kind and each one of them represents a unit of the binary system, now three centuries old.

What Leibniz developed as a pure academic structure of ideas today proves to be the foundation of the most important machine of the age of space travel, the computer.

When Western investigators interpreted Chinese mathematical literature, the word "presumably" cropped up with great frequency in their writings. If they were mathematicians, for better or worse they often had to depend on bad translators, and the results were frequently disastrous. If a sinologist sifted through the writings, the contorted mathematical symbols and operations remained a book with seven seals and the results of his efforts could conduce only to misunderstandings.

Admittedly, Chinese mathematicians did not make it easy for their later interpreters. The symbols were often obscure and by no means uniform. Thus it could happen that two contemporary works dealt with the same mathematical problem but the presentation was entirely different in each case. Many investigators view these complicated and non-uniform symbols as a reason for the fact that Chinese mathematics remained in a blind alley after the Sung period. We must also make mention of another viewpoint here: in the beginning magic and divination surrounded mathematics in China. It was essentially utilized for calendrical calculations. Hence the question naturally arose as to whether the future of individual destinies could not also be determined with its help. The mystical embellishment and the obscure description of mathematical operations, probably deliberate in most cases, compounded and compounds its confusion.

Nevertheless, many findings of the old Chinese masters have emerged from the archives into the light of day. Let us single some of them out in loose sequence. The history of the zero is a good beginning.

The zero designates an empty position within the decimal number. Only through it does calculation in the decimal system become simple. All we need besides it are the numbers 1 to 9 and then we can very easily represent large figures at will.

School children are taught that the zero was discovered in India in the ninth century. Lesser known is the fact that the zero emerged in the Indochina region (between Sumatra and Cambodia) before it was known in India. Could that have been an influence coming from China? The historian will smile commiseratingly and point out that in Chinese literature the zero is identified for the first time in 1247. Nevertheless, the existence of the decimal system can be traced back to the third century in any event. The Chinese mathematicians left a position free precisely where today we expect a zero. This position was frequently marked by a dot. Was this, perhaps, the precursor of the zero? Although we do not wish to belittle the importance of the early Indian algebraists, the fact that the zero appears for the first time on the boundary between the Indian and Chinese culture areas should incline us to reflect on the matter.

To attempt squaring of the circle has become synonymous with tackling insoluble problems. Yet the problem seems so simple: Given a circle, a square is to be constructed, exclusively with the aid of ruler and compass, whose surface is exactly as large as that of the given circle. The procedure amounts to determining whether the number π (*pi*)–the ratio of the circumference to the diameter of a circle–can be represented as a fraction of two whole numbers.

In order to anticipate the final result we shall immediately note here that the German grammar-school master F. Lindemann in 1882 conclusively proved that π is a transcendental number, that is to say, expressible as an infinite, non-periodic decimal fraction.

Since the days of the ancient Egyptians countless mathematicians, lay persons, over-subtle minds, muddle-headed thinkers, technically called circle squarers, have attacked the problem. The cry "Eureka" has resounded over and over again. It rose first of all in the Egyptian Rhind Mathematical Papyrus, written over 3000 years ago, echoing and re-echoing thereafter in the literature of all peoples who concerned themselves at all with geometry. Nobody, in fact, was successful. But this could not restrain new generations from tackling the problem anew with ruler and compass. The number of unsolicited demonstrations so grew that in 1778 the French Academy of Sciences adopted a resolution ordering that communications of this kind be thrown unread into the wastebasket.

The upshot of all these efforts was to push the value ascribed to the number π increasingly higher. As frequently happens the Bible turns out to be an historical source also on this point. In the Old Testament, in the first Book of Kings, Chapter 7, verse 23 we read: "And he made a molten sea ten cubits from the one brim to the other: it was round all about, and its height was five cubits: and a line of thirty cubits did compass it round about." Accordingly, at the time of King Solomon *pi* was equal to 3.

But more accurate values had already been achieved in Babylon. The method of determination is quite simple: the circumference of the circle is approximated by a number of identical triangles. Two sides of a triangle are equal to the radius of the circle, the sum of all the bases of the triangles—the inscribed polygon—then yields the approximation for the circumference. This figure, divided by the diameter of the circle, leads to an approximative value for π. The more triangles there are, the more exact is the approximation. In practice, therefore, it was only a matter of filling the circle with the greatest possible number of triangles of such a kind—a test of patience.

For a long time Archimedes (285-212 B.C.) set the Western standard: his value for π lay between 3 1/7 and 3 10/71.

Mathematicians in China also turned their attention to determining approximations of the value of π, but there are no direct reports on the beginnings. Only one clear indication of such efforts is extant: in 5 A.D. Liu Hsin prepared a standard measure which was a cubic-shaped orifice cut out of the cylinder. From the inscription engraved on it we can conclude that 3.154 was used as the numerical value for π.

The geometer Liu Hui, who lived in the third century A.D., constructed a polygon of 3072 sides and from it determined a value for π ranging between 3.1415927 and 3.1415926. Two hundred years later ten places after the decimal point were known.

This game—after all, for all practical purposes 5 places would suffice—went merrily on. Now it was the turn of Arabs and Westerners. The Dutchman Adrianus Romanus (1561-1615) constructed a polygon with 1073741824 sides (!) and therewith determined π to 15 places after the decimal point. Here the reader may wonder if that is altogether correct. It is not if we calculate that there are around 31.5 million seconds in a year. If Adrianus Romanus had drawn one side every second, he would have required 34 years for it and not had a second's sleep. Perhaps he erred in estimating the number of sides.

Ludolph van Ceulem of Cologne improved the accuracy to 35 places. Consequently, for a long time π in Germany was known as the Ludolphian number.

There was no stopping mathematics buffs after Leibniz published in 1676 an infinite series for the calculation of *pi*:

$$\frac{\pi}{4} = 1 - \frac{1}{3} + \frac{1}{5} - \frac{1}{7} + \frac{1}{9} - \frac{1}{11} + \frac{1}{13} - + \ldots$$

The number of known places after the decimal point rapidly swelled to several hundred. Today, electronic computers can calculate the circumference of a circle of unit diameter with practically optimum accuracy. Is this a waste of valuable computing time? Not at all. The series of the individual figures in π fundamentally exhibits no classification features at all. A series of this kind is called a numerical order of probability. Such numerical orders are frequently required for the interpretation of many scientific experiments.

When the French philosopher and mathematician Blaise Pascal died in 1662, his testament contained a so-called arithmetical triangle. Each number of "Pascal's triangle" amounts to the sum of the two numbers to the right and left directly above it. The reader can easily perceive other regularities. This is no useless game with numbers, but a representation of the principle of binomial coefficients. The figures are required for the solutions of equations of a higher power. It was an important finding of the mathematical science of the seventeenth century.

1
1 1
1 2 1
1 3 3 1
1 4 6 4 1
1 5 10 10 5 1
1 6 15 20 15 6 1
1 7 21 35 35 21 7 1

"Pascal's triangle"

Pascal's contemporaries would have been quite amazed had they been able to take a look at the page reproduced here from the work of Chu Shih-chieh.

Even without a knowledge of the Chinese characters the importance of the arithmetical pyramid would have dawned on them. The book appeared in 1303, a few years after the return of the Venetian Marco Polo from the Middle Kingdom. A person with a knowledge of the Chinese literary language–and such could be found among the China missionaries–would have gathered from the text that Chu Shih-chieh did not lay claim to being its inventor. He referred to a master of the twelfth century, a full five hundred years before Pascal!

Presumably the Jesuits did not see any edition of this standard work in mathematics at the beginning of the seventeenth century while they, as the first Europeans, sojourned for an extended time at the Imperial Court. The Chinese mathematicians had hardly progressed further at the time of

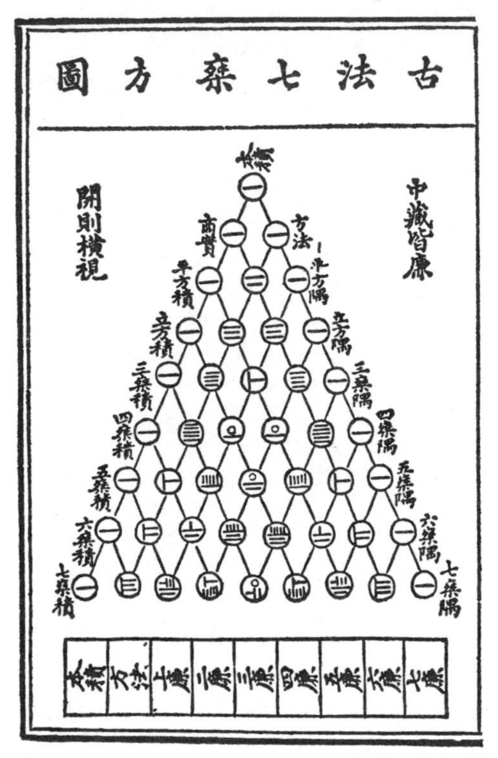

Pascal's Triangle, as published by Chu Shih-chieh in a Chinese book on mathematics (1303). The figures are adaptations of rod numerals.

the Ming dynasty. The closed Chinese Wall allowed no stimulating impulse from the outside to penetrate. For the arch-conservative society of the epoch, mathematics–so serviceable in practice–was a subject with which a scholar of reputation and esteem did not concern himself. A copy of the book with Pascal's triangle appeared in Korea for the first time only as late as 1839.

We could dwell further on mathematics in ancient China but

for fear of wearying the non-mathematically minded reader. Nevertheless, a few paragraphs on the magic square are in order here.

In Albrecht Dürer's famous engraving, "Melancholia" holds the compass in her hand, the ruler not far away. Is she, perhaps, pondering the quadrature of the circle? But the reader need not focus his attention on that. The object of our interest, a magic square, hangs on the wall to the right. No matter how we add up the numbers of four neighboring fields, horizontally, perpendicularly, or diagonally, the result is always 34. Consider that the 16 fields contain all numbers between 1 and 16; no number appears doubled or omitted. This is the oldest known magic square in the West. Soon it formed part of the standard equipment of every astrologer, alchemist, and charlatan mindful of his reputation. Where did this play with numbers start?

Although we find beginnings in the Greece of the first century, the oldest magic square doubtless stems from China. The Lo Shu square is described in the fifth century B.C. Its origin is lost in the mists of mythology. According to the ancient texts it was the gift of a turtle, taken from the River Lo, to the Emperor Yu, and the nine fields correspond to the nine maxims of wisdom that this emperor received from heaven. The investigators did not find it so easy to reconstruct the first magic squares on the basis of the writings that have been handed down. In one passage, for example, the Lo Shu square is described as follows: "Two and four are the shoulders, six and eight the feet, three stands to the left and seven to the right. It wears nine on the head and is shod with one, while five is in the middle."

It required a great deal of acuteness and a gift for combination to recognize a magic square from this description.

Soon magic squares became more comprehensive, first 16 fields, then 25. Each mathematician strove to invent his own square, different from others. This proved not to be so difficult; empirical rules for the construction of squares were soon discovered, and the combinatorial probabilities are many. Thus

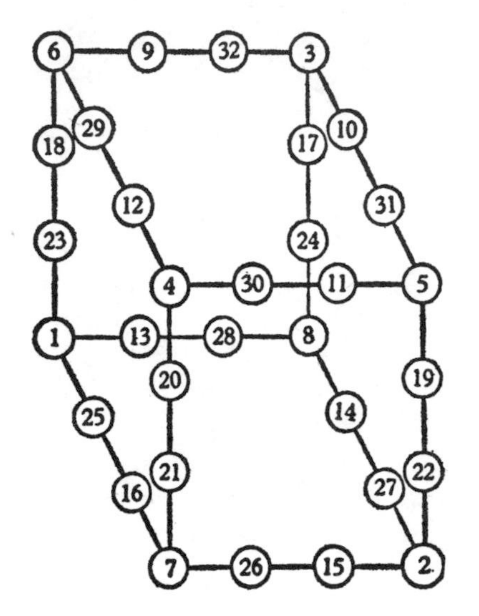

The magic cube of Pao Ch'i-shou (nineteenth century).

880 different magic squares can be composed on the basis of the first 16 numbers. Nor was the construction limited to the square: magic stars, circles, spirals, triangles, and polygons came into being.

This arithmetical acrobatics flourished most impressively in China and Europe in the nineteenth century. The reader is invited to make his own investigation of Pao Ch'i-Shou's three dimensional magic cube reproduced on this page. It can be demonstrated with many examples that Chinese mathematicians, up to the end of the Sung dynasty (middle of the thirteenth century), had no need to fear comparison to their Egyptian, Greek, Arab, and European counterparts. Algebra especially was widely developed. It seems to be characteristic for the course of things in China that mathematics was geared more to practical application. Mathematics for its own sake–as was found, for example, in classical Greece–was hardly pursued. Only seldom do we find strict proofs of mathematical propositions. This is perhaps natural in a country with gigantic engineering constructions such as the Great Wall and the Great Canal. The complicated irrigation system and fiscal computation in a centrally administered country also fostered applied mathematics.

We can only speculate on the reasons why the otherwise very promising development largely came to nought after the Sung dynasty. Was it due to the non-uniform, complicated notation? Or to the social structure under the Ming dynasty?

The Chinese Wall was wide open at the time of the Mongol rule (Yuan dynasty). For the last time a lively exchange took place with Arabic and Persian mathematicians. When the Ming dynasty came to power in the fourteenth century, it immediately established a conservative hierarchy which looked down on science. It viewed science as a form of necessary manual labor. But a gentleman of standing did not concern himself with it. The cultured man had exclusively literary interests. At the same time, the Chinese Wall debarred influences and impulses from the world of the "barbarians." The portals were closed and one stewed, so to speak, in one's own juice. In this period we recognize for the first time that the closed Chinese Wall had a clearly negative effect on China herself. There had already been periods before, characterized by the closed Chinese Wall, but the disadvantaged ones were those living outside it.

Neither the rulers of the Ming dynasty nor those of the Manchu perceived the baleful influence of the self-encapsulation which now hindered further development. When this realization finally dawned on them, it was too late. China, meanwhile, lay very far behind the industrialized nations of the world. Gigantic efforts were required to close this gap.

3.

Maps on Silk, Paper, and Stone

"HERE I am, O Crown Prince." So saying, Ching Kho crouched at the feet of his lord in the imperial palace of Yen and awaited orders.

"I am satisfied with your services and assuredly I am thankful to you for them. I have one further order after which you may retire to your estates in the province as has always been your wish, Ching Kho." A barely perceptible passionate undertone now crept into the voice of the Prince Yen Tan Tzu. "The king of Ch'in is casting covetous eyes upon my kingdom. He is strong and we are weak. Betake yourself there as my ambassador and assuage his anger. You are not going there empty-handed." Thereupon Yen Tan Tzu clapped his hands, and a page appeared with a closed basket.

"Tsin Wu-Yang will accompany you and present my gift."

With these words he drew back the covering. Ching Kho stood there, unflinching. After all he had seen severed heads often enough before.

"It is the head of Fan Yu-Tsi. The sight of it will gladden the king of Ch'in. Then you are to give him this." The crown prince removed the silk cloth from a long wooden case.

"Open it and look at my present for the king of Ch'in! A map of the border region of Tu-Khang is drawn on the scroll. He will study this map with interest, since I am giving him this land as a gift. His attention will be momentarily distracted from you. Then act!"

Ching Kho's look fastened unwaveringly on the case. The silk scroll inside it hardly interested him; nor was he impressed by the magnificent carving and the inlaid characters. His attention focused on the handle that projected from the silk—the handle of a dagger! Betraying absolutely no emotion, he bowed low, almost touching the floor with his head and answered: "I hear and I obey."

The "ambassador" then withdrew, walking backwards. Yen Tan Tzu's eyes followed him with an air of evident satisfaction: Ching Kho, after all, was an experienced agent, the problem was solved.

This would have been a simple and swift solution. The kingdom of Ch'in had been constantly expanding for generations. Neither the nominal overlords of China, the Chou kings, nor the neighboring kingdom could block this expansionism.

Not without reason was it called the period of the Warring States. In the beginning each one fought against the other, and alliances changed from one season to the next. But soon a clear trend was discernible: Ch'in was swallowing one rival after the other. This was not accidental. The rulers had deliberately built up a centralized administration in the territory lying in northwest China. The king held all power in his hands and controlled even the farthermost corners of the realm with the help of his superbly trained officialdom. The ruler no longer needed to fear that rivals in his own country would dispute possession of the throne. The nobility was no more. He could give his undivided attention to the inner consolidation of the country. He erected a wall against the nomads in the north, while his well-trained and hard-hitting army restrained the aggressive lusts of the rest of the neighboring states. Thus peace was secured within and without. The irrigation system was improved, the spring floods of the tributaries of the Wei River flowed into well-built canals and reservoirs thanks to a vigorous program of public works. The peasants regularly obtained good harvests and thereby strengthened the economic backbone of the State. An extensive network of roads spanned the country, allowing for the rapid transport of goods and troops.

An extraordinarily powerful empire–economically and politically–arose in two generations; it enjoyed great inner stability as well. Its rulers began to look beyond the borders. The neighboring kingdoms were weak. They had neglected their agriculture in consequence of ceaseless wars and possessed no reserves of any kind to withstand long campaigns. They were an easy prey for the Ch'in. The conquered areas were not only annexed, they were fully absorbed by the Ch'in. Neither the ruling families nor the individual features of these kingdoms survived, all and everything had to adjust to the strict rules of the victor.

Even the hastily formed alliances broke down under the onslaught of the Ch'in, whose armies were not only well-trained but who could depend on a steady flow of abundant supplies from the homeland. This was made possible by a perfect organization.

Around 230 B.C. the kingdom of Yen was still the only independent realm in China, and it saw the handwriting on the wall. Too weak to place its trust in the outcome of the ineluctable armed conflict, the rulers of Yen had recourse to another means: assassination. The moment seemed opportune. A general of the Ch'in, Fan Yun-Tsi, had deserted and found political asylum in Yen. It would certainly delight the king of Ch'in to receive the general's head. The only hitch was that the Crown Prince of Yen was a friend of Fan Yun-Tsi. He begged his friend to leave the country so that no one might yield to the temptation of profiting from the death of a refugee seeking protection. When the general learned the reasons for the plea and heard of the planned assassination he killed himself on the spot. His last words were: "If my head can contribute to the destruction of tyrants, I shall die with great joy."

That was how the head of the defecting general ended up in the basket and set out on the journey to Ch'in. Yen Tan Tzu believed that by this ruse he had saved the kingdom of Yen.

How changed the expression on his face would have been had he been able to see just two weeks ahead and witness the dramatic scene that unfolded in the reception hall of his enemy: the dagger, quivering, was stuck in the base of a wooden column. The man for whom the dagger had been intended had nimbly leaped to one side, exhibiting a great presence of mind. The silk map had slipped out of his hands. The page Tsin Wu-Yang cowered on the floor, screaming, glancing sideways with panic-stricken eyes. The kingdom's physician had grabbed Ching Kho. The king's guards had already stormed into the chamber, the mission was a failure.

This historical event took place in 227 B.C. In itself it would command no special interest today, since scenes of this kind were of frequent occurrence in the palaces, had it not been for a trifling detail: the map. It is the first clear, unambiguous, and datable reference to a map in China. There had already been geographical representations in the Middle Kingdom. The first were presumably maps of the Hwang Ho (Yellow River), which played a prominent role in the life of the Chinese. In good times it irrigated and fertilized the country and became the nourisher of men. At times, however, it burst the dikes and then disaster and misfortune broke over a great part of the population. The people made propitiatory sacrifice to the river and sank boards with drawings into the stream in order to show it the direction along which it should flow.

Life in the Babylonian kingdom was also based on a river culture. Thus it is not particularly surprising that the first maps out of Babylonia represent the two dispensers of life, the Tigris and the Euphrates. These maps engraved in sun-dried brick give a picture of the irrigation system in the year 3000 B.C.

Chinese annals also contain narratives describing how maps of rivers were given to the emperor to take along with him to his grave. Perhaps with their help the rulers wanted to attest their good deeds on earth since the regulation of rivers was always their primary task and obligation.

Maps of larger territories probably first appeared when the individual regions of China were united into a single kingdom. Texts tell of maps of nine provinces at the time of the Shang dynasty (1520-1030 B.C.), none of which has survived. Professor Albert Herrmann, however, has shown how we can eliminate misunderstandings and forgeries from the old reports and reconstruct the maps to which the texts refer. He succeeded in redrawing the map of the last king of the Shang dynasty, and his portrayal of this effort is as suspenseful as a classic detective story. Under the Chou dynasty (1030-722 B.C.) there were special officials whose job was to keep geographical knowledge up to date, and this was of great value to researchers. These officials had the lists of the Shang at their disposal and their commentaries on them have been preserved.

Here we must mention a peculiarity of Chinese maps: they represent the earth as square. There are philosophical reasons for this: the opposition between heaven and earth was expressed in the Chinese world view by ascribing to heaven, symbol of bright, warming Yang, the shape of a hemisphere, and to earth, symbol of dark, cold Yin, the shape of a square. The Son of Heaven (the emperor) has his residence in the center of the world. He is appointed as mediator between heaven and earth and is responsible for their reciprocal harmony. If he is derelict in the performance of his appointed duty, the mandate can be withdrawn from him. Nevertheless, the emperor did not function as a kind of high priest; the task of achieving and maintaining harmony devolved only upon his statecraft.

On the basis of this world view, presumably developed at the time of the Shang dynasty, it is at once comprehensible and necessary that the capital should have a quadratic form. This is in utter contrast to other countries, where the royal residence is adapted to natural features of the landscape, nestling in the bend of a protecting river, for example.

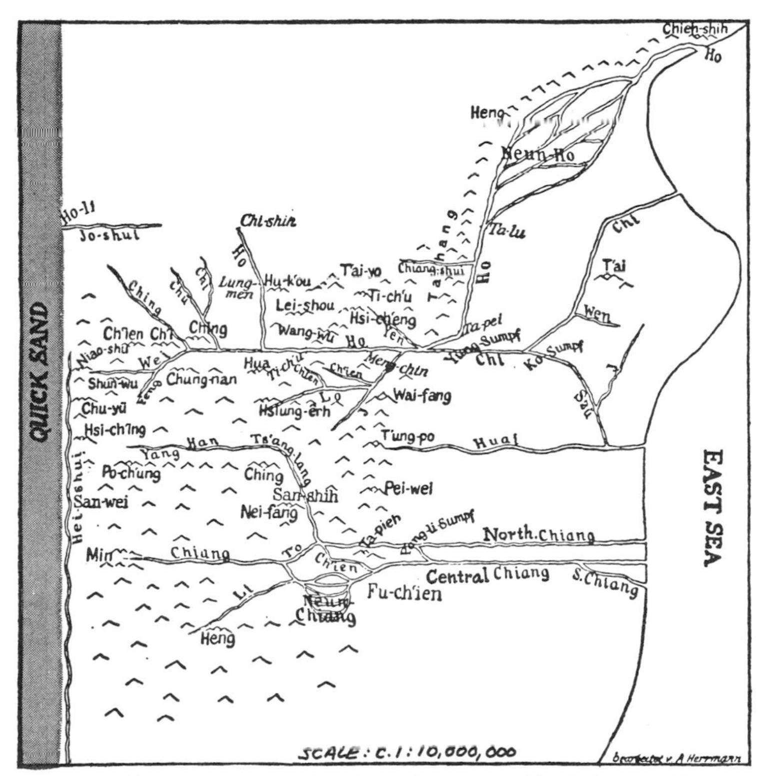

Copy of a Chinese map from the Shang Dynasty (1520-1030 B.C.). The map shows the region between Longitude 110°-120° E and Latitude 30°-40° N. The great loop of the Yellow River is not represented.

Hence the conventional Chinese map is a sequence of concentric squares. The imperial palace is in the middle, and the lands of the barbarians, washed by the seas, lie wholly outside it. Civilized people lived only within the three innermost squares. And the Chinese Wall marked them off from the regions inhabited by barbarians.

Confucius taught this world view and therefore we find no trace of the curvature of the earth's surface even in maps of

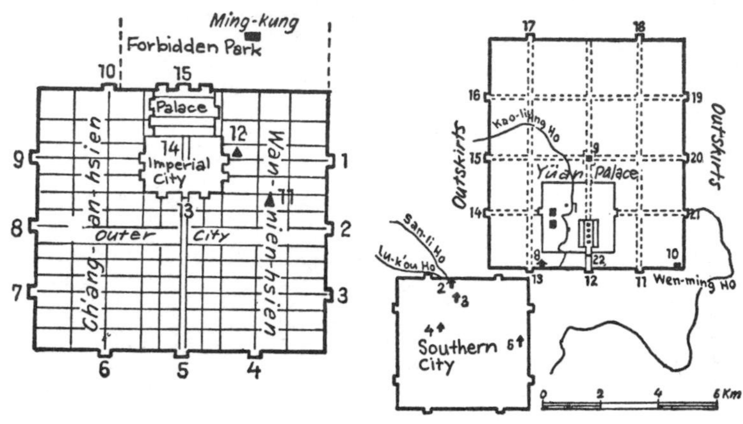

The imperial palace as the center of the world. Since the Chinese conceived the earth as a square, the residence had to be constructed in the same shape. Left: Sian, at the time of the Sui and Tang Dynasties (581-906). Right: Chanbalic, built by the Mongols, 1267-1271.

1200 A.D; which are otherwise so modern. However, some scholars doubted this orthodox cosmological world view. Although their opinion had little influence on the hierarchy dominated by the teaching of Confucius, it has come down to us: "Heaven is like an egg, earth swims in the midst of it like the yolk. Heaven is vast and earth is small" (c. 300 A.D.).

But let us return now to the Ch'in dynasty, which ultimately was to give its name to the Middle Kingdom. Six years after the abortive *attentat*, the king of Ch'in conquered the remaining territories in China and annexed them to the Middle Kingdom. He himself ascended the throne as Shih Hwang-Ti. He assembled all available maps in his capital, Sian. The maps of the individual provinces incised on wooden boards could be arranged alongside each other to form a whole map. He must surely have spent many hours studying it in order to plan new campaigns and to fix the course of the Great Wall.

His reign was short. After his death in 207 B.C., the soldiers

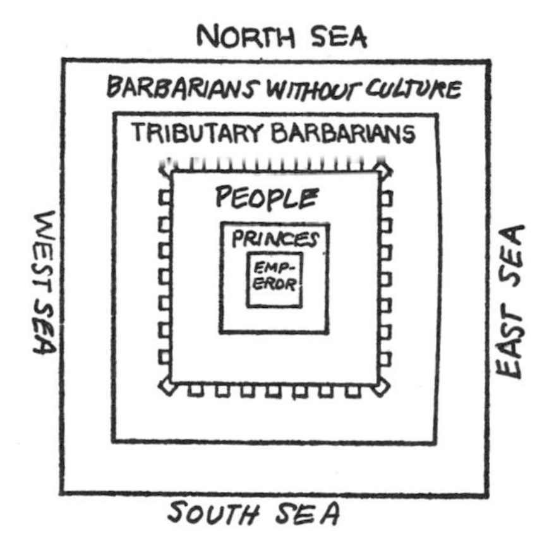

The orthodox world view of the Chinese. Human beings lived only inside the Great Wall. In the center the Emperor attended to the key task of preserving harmony between heaven and earth.

of the Han dynasty conquered the capital. The victorious army gave itself over to the pleasures of pillage. While the officers and generals personally enriched themselves in the palaces of the foe, a mode of conduct that by no means was either unusual or unseemly, the commander-in-chief laid the foundations for the long rule of the Han: he was not interested in either gold or silver. Instead, he searched through the imperial chancelleries for the maps of the Ch'in dynasty. Before the enormous rooms went up in flames, he had managed to bring all such documents to safety.

A farsighted man of this kind was not always present when war again ravaged the country. Consequently, many of the old maps have not survived.

Naturally these were maps of all parts of the country in the rigidly organized Han kingdom where the centralized bureaucratic State was brought to perfection. The officials prepared such maps for their administrative districts, devoting particular attention to the cities. They completed the maps of the Ch'in dynasty and transferred them onto silk and later also onto paper. Few of these have survived and come down to us. The only one extant today is a city map of Sian from the end of the third century A.D.

Thanks to Phei Hsiu (224-271 A.D.), called the father of Chinese cartography, we know a great deal about the art of map making. He gathered all available data, sifted them critically, standardizing the representation of cities, mountains, rivers, and landscapes, and made a new map of the Middle Kingdom.

In the foreword to this atlas he laid down all the rules, which a modern cartographer could also heed. He was the first mapmaker in China to use a grid system. The regions represented are overlaid with a rectangular system of coordinates. Was that his own idea?

Almost five hundred years before him Eratosthenes of Cyrene had introduced the measurements of longitude and latitude, whereby he already took into consideration the curvature of the earth's surface. A great many indirect contacts between China and the countries of the Mediterranean region took place by way of Roman Syria and the Old Silk Road. It is not even necessary that Phei Hsiu himself possessed a map with a grid system from the Hellenic world. The report of a traveling merchant who had here and there seen a map overlaid with squares would have sufficed. The advantages of such a system of subdivision would not have escaped the cartographer. The assumption of an indirect transmission of this idea is also supported by the fact that Phei Hsiu really applied rectangular coordinates. If he had possessed Ptolemy's world map, perhaps he would have given some consideration to the curvature of the earth's surface.

This need not necessarily have followed, however. When 1600 years later Canada was exactly surveyed, the surveyor's office covered the gigantic territory with a grid system. It consisted of mile-long squares, thus disregarding the earth's curvature for simplicity's sake. To this day all roads in central and western Canada run very nearly on the grid.

Phei Hsiu's Map of the Kingdom and the Barbarian Countries comprises 18 sheets. But since the preface and index map take up one sheet each, the actual map was made in 16 sheets. The father of Chinese cartography also represented the earth as a square. He oriented all maps on the north as is still customary today. The scale was 1:5 million. Unfortunately, this highly instructive work is also lost. Only a pale reflection of it has come down to us. In 1701 an official copied the index map to the whole work, presumably erroneously and in a stylized way, as was customary at that time.

If Phei Hsiu was able to draw very detailed maps, and this is confirmed by the Chronicles, it must also have been possible to measure distances quickly and reliably. His contemporary, the engineer Ma Chun, either invented a hodometer or, perhaps, improved a primitive precursor of the instrument. The notion of a hodometer suggests itself by simply observing a wagon in motion: each turn of the wheel corresponds to a definite distance traveled. The number of turns can easily be measured with the help of a simple arrangement of cogs. If the circumference of the wheel is known, it is easy to calculate the distance. Every automobile mileage meter functions this way.

A chronicle from the year 1023 exactly describes Ma Chun's hodometer. In the 1930's a Chinese scholar reconstructed the instrument according to these data; it functioned satisfactorily. An acoustic signal indicated the distance traversed. At the completion of every li (about four-tenths of a mile, but this figure varies), the cogs guided the arm of a mechanical wooden figure that, in turn, struck a drum. After every ten li, another figure gave a roll on the drum.

Is it an accident that Ma Chun's hodometer measures distances by mechanical musicians? Presumably not. When the emperor traveled in the country a carriage with musicians always formed part of his entourage. Their job was not only to gladden the ruler with music but to mark the distance traversed. A blow on the drum was a signal to the emperor that he had come a li nearer to the goal of the journey. A painting of this carriage for musicians from the first century of the Christian era has been preserved; it bears a great resemblance to Ma Chun's hodometer. Who knows? Perhaps the musicians depicted were not flesh and blood characters but puppets. In view of the manual dexterity of the Chinese and their predilection for mechanical toys this would not be surprising. The painting gives no clear answer to this question.

The same method, namely of quickly and effortlessly determining distances traversed with the help of the rotating wheel, is already described in the first century A.D. by the Greek savant Heron of Alexandria in his work *Dioptra*. Presumably it is not his own invention; he was probably merely passing on knowledge of an older model. Apart from a brief mention in Roman times, however, the knowledge of the hodometer was soon lost in the West. It was not rediscovered until the Renaissance when Leonardo da Vinci made a design that was then used for many years.

But let us return now to the cartographers of China in the third century A.D. The officials of the Imperial Chancellery for Countries and Provinces were constantly in search of new methods that could make valuable maps immune to destructive influences. At the same time maps always had to be light and manageable. For every time the ruler set out on a tour of inspection, the director of the Chancellery had to accompany him and comment on the features of the landscape as they passed. Nor did the generals find a map carved in stone very

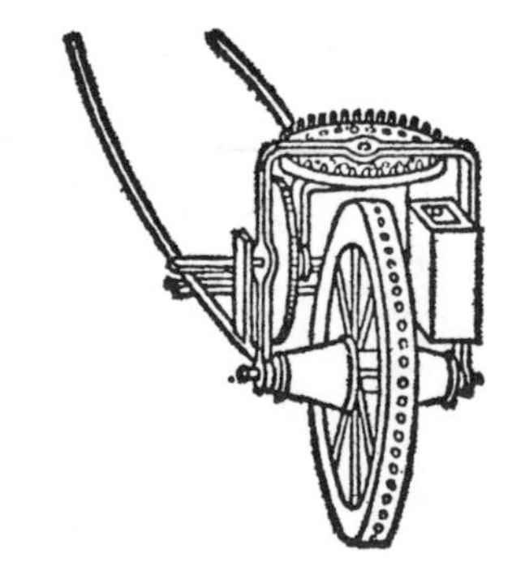

Reconstruction of a hodometer of Heron of Alexandria (100 A.D.). Each time the wheel completed a certain number of revolutions, the mechanism deposited a pebble into the receptacle attached to the side.

convenient. The baggage train was heavy enough. Hence for a very long time silk remained the material on which elaborated contours and topographical symbols were inscribed. Paper was first considered as an alternative in the second century.

But Chinese cartographers in the third century A.D. fancied they had found an even better method. Here is what the Annals of Shih I Chi (Remembrances of Neglected Matters) says about it: "Sun Ch'üan searched for a skillful painter who could draw a map with mountains, rivers, and other landscape features for military purposes. The younger sister of the Chancellor was recommended to him and he instructed her to draw the mountains, rivers, and lakes of the new provinces. She suggested that the map be embroidered instead, since the colors of a drawing would fade in the course of time. And this was done." The remarkable map must have been lost during the military expedition, for we hear nothing more concerning this interesting solution.

Cartography in the West achieved its peak with Ptolemy's world map, after which it went into a rapid decline. In the course of time the topographical representation of the world was placed in the straitjacket of theology until well beyond the turn of the millennium. The earth was depicted as a flat disk which is subdivided in a symbolic way. There is no discernible connection with reality on these "maps." To be sure, in the Far East Confucius taught that the earth was a square, but this could not prevent Chinese cartographers from representing reality within the square. They continually gathered further knowledge concerning the Middle Kingdom and the "barbaric" countries bordering it. They corrected errors, standardized the names of cities, mountains, and rivers, elaborated the reports of military expeditions to distant regions, and sent out their own teams of surveyors to study the headwaters of important rivers.

Chinese geographers sought for the source of the Yellow River with the same passion exhibited by their colleagues, many thousands of miles to the west, with respect to the source of the Nile. Attempts have been made since time immemorial to locate the source of the Hwang Ho. The fields of its catchment would be without fertilizer were it not for the masses of silt it carries along with it on its course. Like the Nile in Egypt, the Hwang Ho was the source of all life in northern China.

The upper course of the Yellow River is represented for the first time on the map of the Ch'in dynasty. Nevertheless, what Herodotus said about the source of the Nile applies equally to the Yellow River: "Of the sources of the Nile no one can give any account . . . It enters Egypt from the parts beyond."

In the Far East the solution of the riddle seems to have been found earlier: in 635 A.D. a general of the Tang period was pursuing rebellious tribes far into the Tibetan highland. There at a height of about 13,500 feet he found the source–or so he believed.

In 1858 the English explorers Richard Burton and John Speke identified Victoria Lake as the source of the Nile, yet it was not until 1875 that Henry Morton Stanley pinpointed Ripon Falls as the actual source of the Nile.

The Yellow River starts in regions so remote that it was not until 1953 that the beginning point of the second largest river in China could definitely be established–and, for that matter, quite close to the point fixed by the Tang General named Hou Chun-Chi in 635 A.D.

The Nile and the Hwang Ho were not only of similar importance to the dwellers on their respective banks, but the history of their discovery also exhibits many parallels.

In 785 Chin Tan received the imperial commission to collect the geographical cognitions of the time and to prepare a map of the empire of the Tang dynasty. The work was completed after sixteen years, the "Map of All Chinese and Barbarians Within the Four Seas." A tremendous accomplishment indeed: it was 30 feet long and 33 feet high with geographical details, all arranged within a system of coordinates. One inch on the map corresponded to 100 li in nature. All Asia was encompassed. Unfortunately, even this remarkable cartographical achievement fell victim to the ravages of time.

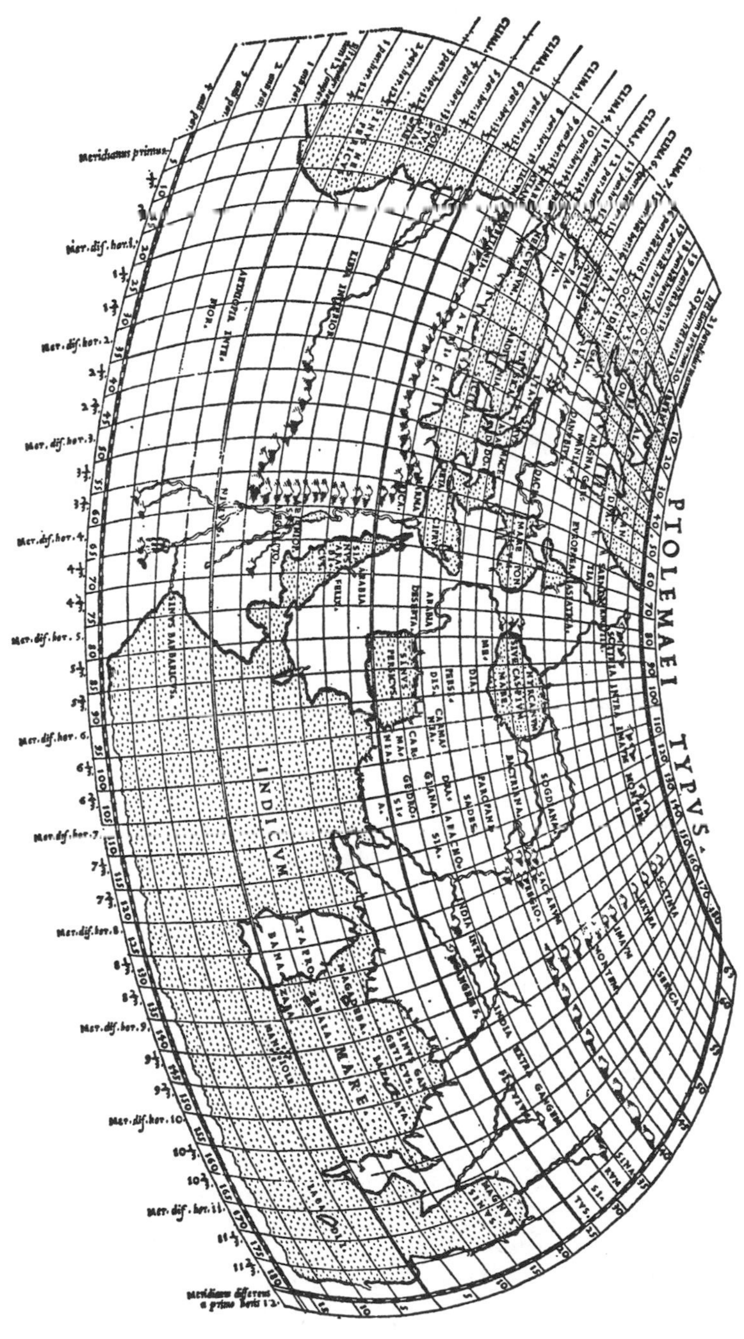

The world as seen by Ptolemy.

On the other hand, two documents made of a more durable material–they are carved in stone–are preserved in Sian today. This was a favorite method employed at this time, since star maps were also fixed in this way. Not only are such maps almost indestructible, but casts can be made from them and thus many copies are simply and easily obtainable.

There are two maps of China from the middle of the eleventh century. The "Map of the Tracks of Yu the Great" especially is drawn so exquisitely with such a love of detail that up to then it had no match anywhere else in the world. The historian can compare the depicted river systems of the Hwang Ho and of the Yangtsze Kiang to those drawn almost a thousand years earlier. Chinese cartography thereby achieved a peak of perfection which the West was not to surpass until the end of the seventeenth century. The officials of the Chinese court, of course, despised foreigners as barbarians–at least in the times of the closed Chinese Wall. Nevertheless, the reports of these foreigners were recorded and carefully evaluated by the officials appointed for that purpose. The observations of Mohammedan travelers were of special interest. The Mongolian horsemen penetrated deeply into Europe and came back with firsthand geographical knowledge. Maps, of course, played an important role in the plans for conquest. The maps had to be accurate in order to coordinate the movements of the individual armies successfully.

There was a rich accumulation of geographical knowledge in the fourteenth century and the need for a map of his kingdom must have likewise been felt by Genghis Kahn. A map of the Mongolian empire was practically identical with the known world at that time. So it is altogether logical that in 1320 a map came into being which encompassed the known world between the Azores and Korea. The Arabic name for Germany, Aleimania, also appears on the new map, marked phonetically in Chinese characters. Discernible here is the influence of the Moroccan Mohammed Al-Idrisi (1099-1164), whose world map, which was fifteen years in the making, presumably also reached China. But whence derives the knowledge of the

Azores? This group of islands, which had been well known in ancient times, had long been forgotten again. It was only at the end of the fourteenth century that the Portuguese discovered them–and kept it a secret! Africa too, in its triangular form, is correctly represented, the tip correctly pointing south. A multi-storied pagoda on the Mediterranean coast marks the lighthouse of Pharos. The most important city of Europe is indicated in the region where Budapest is located today. Was this perhaps a reference to the residence of Attila? Several other riddles presented by the map await their solution, since up to now only the Asiatic part has been explicated. A copy of this map, dating from the fifteenth century, is preserved in Japan. Unfortunately, the original, the most reliable world map created up to that time, has been lost.

The expeditions of Admiral Cheng Ho, undertaken between 1405 and 1431, again broadened geographical knowledge. A series of special maps of particular interest to mariners was produced. The admiral's expeditions, which included up to 100 ships and 50,000 men, did not so much serve the purpose of acquiring geographical knowledge–that was only a by-product as it were–than that of tracking down the defeated Emperor of the Yüan dynasty. He was supposed to have escaped by sea.

At this point let us drop the curtain on the Chinese stage and briefly recapitulate the cartographic high points of the West. Two figures from antiquity dominate the stage. First of all there was the Greek Eratosthenes of Cyrene (276-194 B.C.) who for a long time was the director of the famous library of Alexandria. He introduced longitude and latitude, viewed the earth as a self-evident sphere, and determined its radius with such an accuracy that the measured value deviates only about 50 miles from the true measurements. Later, the Alexandrian Ptolemy (87-150 A.D.) summarized the geographical knowledge of his time in a world map.

His authority was uncontested in Europe until far into the Middle Ages, even though his world map was lost. What followed subsequently no longer had anything to do with the scientific mode of working that characterized the ancient world.

The earth again became flat and constrained into a form that evidenced religious zeal rather than knowledge. The beginnings in the direction of a well-founded cartography began only in the thirteenth century.

Its development was stimulated by the then beginning sea trade beyond the borders of the Mediterranean. The expedition sent out by Henry the Navigator to find the sea route to India for Portugal especially contributed to this development. The old representation of the world as drawn on the map of that time was an obstacle in the age of mercantilism–it had to give way. The new geographical representations, which established links with the old Greek traditions, were a visible expression of this. Another three centuries went by before the Venetian Girolano Ruscelli succeeded in making a usable reconstruction of Ptolemy's world map. Thereafter cartography made giant strides forward with which neither the Arab nor the Chinese world could keep up.

And today? Stretches of land are measured from an airplane and mapped with an accuracy of which the ancients never dreamed. The automatic aerial photograph interpreter does the work of a whole army of geometers and completes it in a few days.

A few photographs taken from artificial satellites suffice to gather the material for a world map. Weather satellites circle the earth at great altitudes. The photographic information stored on magnetic video tape is transmitted to ground stations, interpreted by computers, and translated into world-wide weather charts.

No region, no matter how inaccessible, remains unsurveyed, the secret satellites photograph even the remotest corners of the world. Nevertheless, the cartographic knowledge thus acquired is not always destined for the public; as military maps they are kept under lock and key and stamped "top secret."

This is hardly new, it has always been thus. As witness let us cite the well-known chronicles of Shen Kua from the eleventh century. "In the time of the reign of Hsi-Ming, ambassadors from Korea came to bring tribute. In every large city on their way they asked for the local maps. These were prepared and given to them. Mountains and rivers, roads, forts, and escarpments–nothing was hidden. They conducted themselves similarly in Thiehchow. But Ch'en Hsiu, the prefect of Yangchow, outsmarted them. He asked them to give him all the maps under the pretext that he wished to make copies of them. Hardly were the maps in his hand, when he burned them all and wrote a detailed report on the whole affair to the Emperor."

4.

The Great Wall

EVERYTHING was suffused in a reddish light. The extremely fine particles of dust burned their eyes, and they turned their faces away from the blazing hot wind blowing in from the Gobi Desert. The workers no longer saw the steed, nor did they still hear the beat of its hooves. What else remained for them to do but to settle down in the protection of the finished section and cook a meal?

The wind subsided, the finest grains of sand sank back into the earth, and now their vision was unimpeded in the endless desert. The workers energetically fell to the task at hand, brick walls rose up to the left and right, and were filled with earth, fascines, and gravel. The Wall grew steadily longer. To be sure the magic steed of the Emperor Shih Huang Ti had disappeared from view, but all the builders had to do was to keep to the old direction. Time passed; the foreman grew uneasy and sent out a scout. Where were the Emperor and his horse? The scout came back from a direction different from that which had been expected. He had found the two far away and in the direction of the sunrise. The workers had to go back to the old resting place where the track had been lost in the sandstorm–some 20 miles.

Meanwhile Shih Huang Ti, the Great Unifier of the Empire, raced tirelessly on. The Wall arose on the track of the magic steed. Wherever it had stamped its heels, a watchtower would grow out of the ground. Impregnable mountains? The magic whip would scare them to one side. The ride went from the Tarim basin in the remote far west up to the Yellow Sea. Between these two points arose the Wall which separated the celestial kingdom from the barbarians. Thus legend has it and also provides proof: the sections branching out from the main Wall which have no discernible purpose.

The fantasy of the people attributes to the just Emperor the fame of being the builder of the longest rampart in the world. But did it really serve as a defense of the Empire which had been united for the first time precisely then? In the year 221 B.C. the kingdom of the Ch'in had incorporated all the other kingdoms between the Hwang Ho (Yellow River) and the Yangtsze Kiang. The conqueror created for himself the imperial title Huang Ti, the Illustrious Emperor. No foe was a match for his armored-car troops and their iron weapons. Such weaponry, of course, made no impression on the wild horsemen in the north and northwest, but no danger threatened from them because they had again fallen to quarreling and were busily fighting each other. Why, then, did the ruler on the Dragon Throne give his ablest general and engineer Meng T'ien the order: "Take three hundred thousand men and build a wall in the North"?

Military reasons played a subordinate role, the armies of Ch'in came back, triumphantly, from all fronts. But now they were soldiers with nothing to do. There were no longer any enemies worthy of mention, so they encamped idly in the garrisons. A dangerous situation. Restless minds–after all, the feudal nobility was not wholly destroyed–could incite the soldiers and lead them against their Emperor. His harsh and distrustful character had indeed already offended many people. Hard work would dispel their rebellious thoughts! The ruler also thought of his subjects in the northern provinces. There the soil was not as fertile as in the valleys of the Yellow River and of the well-watered Yangtsze Kiang. In dry years the crops withered in the fields and the peasants thought enviously of the cattle-raising nomads who always found pasture land for their herds somewhere in the steppes and who never died of hunger. At such times those wretched peasants soon became predatory nomads and descended upon the fertile territories like locusts. It was necessary to keep them within the confines of the Middle Kingdom. The Wall became a visible symbol: China extends to this point; beyond the battlements begins the world of the barbarians. And the Emperor defined the qualifications for membership on either side.

The nomads wandering beyond the Wall had once also been inhabitants of the fertile plains and valleys of China. Many generations before, stronger tribes had driven their forefathers into the economically inferior regions. Soil-tilling peasants became roving nomads, not vice versa.

Seven years later–in the West, Rome had just recovered from the shattering defeat at Cannae–Meng T'ien reported to his sovereign: "The Wall is completed. One foot is washed by the sea, the other is anchored in the western desert."

The construction extended over a distance of some 1500 miles in a generally east-west direction. The base was from 15 to 30 feet wide and the cremallated wall tapered to a width of 16 feet. Towers, about 30 feet high, were constructed at double bow-shot distances atop the Wall whose height ranged from 15 to 50 feet.

Bastions projected from the Wall on the side turned to the enemy: it was a free-fire zone for the archers and crossbow men. The towers not only served for direct defense. News was also transmitted through them with lightning speed, from one point to the other, with smoke signals by day and fire signals by night. Thus the Chinese in the fortified camps and garrisons of the hinterland were always informed about the enemy's movements. Isolated watchtowers in the perimeter also served as observations posts. These independent fortresses, well-stocked with a four-month supply of provisions, located on mountain tops and on the edge of defiles, on passes and on plains, were ever on the lookout for nomads. They signaled the first news, began the attack thrust, and tied down considerable enemy forces during the siege. The building material always came from the environs, but the mode of construction was always the same. Every 26 feet the workers built a double wall, and the space in between was filled with earth, stones, and sand. The most solid fortresses are located in the northeast, where the rock of the mountain range provided a building material of excellent quality. On the plains and in the deserts the workers had to make do with baked brick or simply with solidly pounded earth and fascines.

Is it possible that Meng T'ien with his 300,000 laborers required only seven years to build the Wall? At any rate, that is what the old chronicles report. We must take into consideration the fact that long stretches of the Wall were already in existence before Meng T'ien began his operation. They were, in part, 300 years old and stemmed from the period of the Warring States. At this time many kingdoms, such as, for example, Tsin, Chao, Yen, Wei, and still others fought for hegemony. They erected walls against each other and also as a defense against the nomads. It was a turbulent time with constantly changing fronts. The Persian wars were raging in Greece at this time. After Shih Huang Ti had completed the work of his forefathers and subjected the warring kingdoms to his central authority, all he had to do was to close the gap between the extant bulwarks.

For this purpose he did not even need all the troops who, in the absence of a foe worthy of the name, crouched idly under their tents. In addition to those occupied in building the Wall, enough of a labor force remained to continue the construction of canals already begun in the sixth century B.C. Soon a through-waterway connected the Yangtsze Kiang and the Yellow River. In the moist south the peasants garnered up to three harvests per year. They paid their taxes in natural

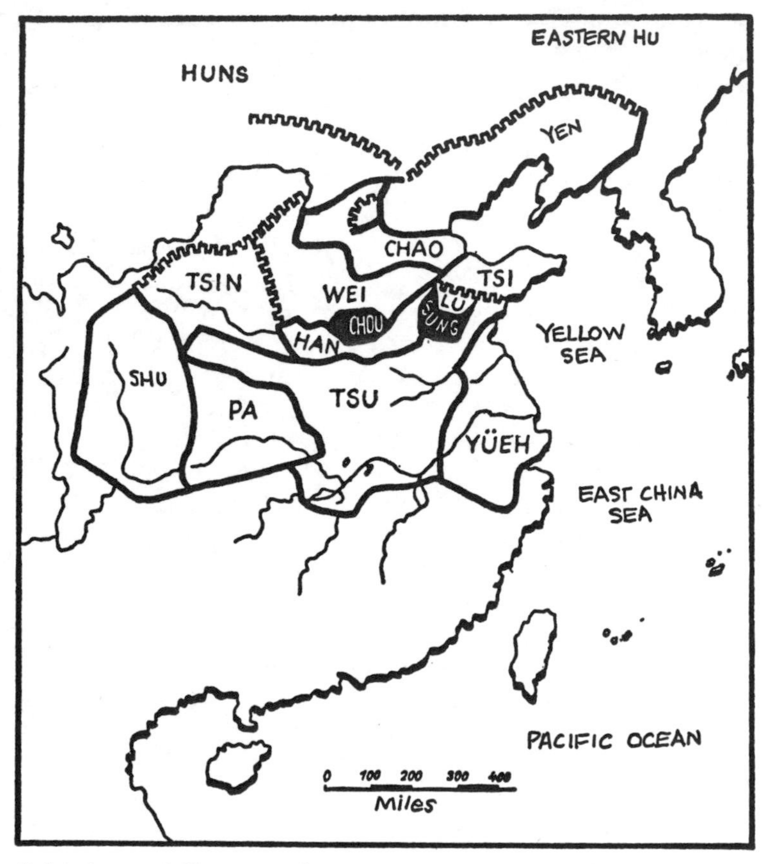

Political map of China around 300 B.C. Parts of the Great Wall had already been built.

produce, and so an endless chain of junks made its way along the Great Canal into the strategically important northern part of the country. Grains were stored in government granaries, which sold them at low prices in lean years and bought them in years of surplus. So there was little opportunity for speculating in grain. The waterways were relatively safe, but during the overland journeys to the construction sites on the Wall many provisions were lost. The ox carts traveled thither very slowly, and the drivers had to eat. Some sold their cargo en route at their own risk, while others lost their carloads and perhaps even their lives to roving bandits. Frequently enough only one of a hundred sacks shipped arrived at the destination. But no matter how great the problem of supply may have been, the fact remains that the Wall was finished after only seven years.

The country's continuous exertion in community projects was not without consequences. The Great Wall, the Great Canal, an extensive network of roads, costly wars with expeditions as far as North Vietnam, all undertakings of a kind that individually would have strained an already powerful empire, exhausted the first imperial kingdom. Unrest broke out when the restless ruler died in the year 210 and a weak-willed son followed him on

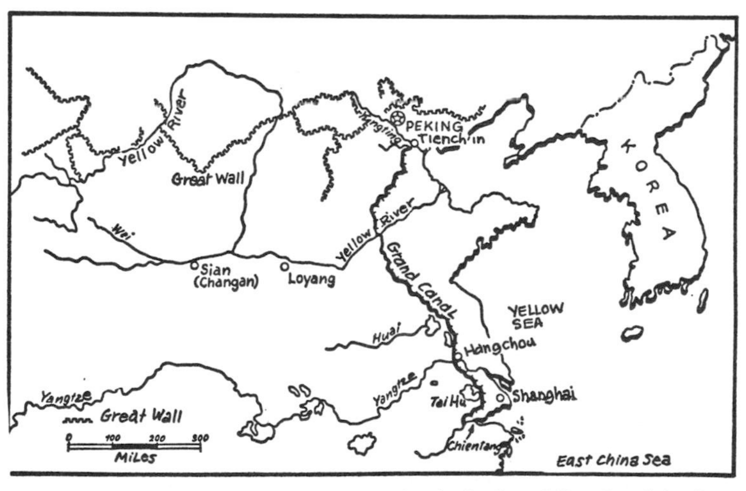

The two greatest ingenious stuctures of mankind: the Grand Canal and the Great Wall.

the Dragon Throne. This marked the end of the Ch'in dynasty. Its memory continues in the name China.

It is worth our while to dwell a bit longer on the architect of the Great Wall. Meng T'ien was not only a successful general and an excellent engineer but also a master of the pen, or more exactly of the writing brush. According to popular legend he was the inventor of this fine instrument of artistic calligraphy. Probably he merely improved the brush; still, this alone shows the contrast in his character; concern with something so fragile, so delicate as the writing brush, while at the same time he was building the longest defensive fortification in the world. None of his literary creations has survived. Archimedes, another engineer of antiquity, still very famous today and separated from Meng T'ien by the expanse of a continent, was not only his contemporary. He, too, suffered a violent end: Archimedes died by the sword of a Roman soldier. The successor of Emperor Shih Huang Ti sent the engineer the Silk Cord, an invitation to self-strangulation. Why? The eunuchs had gathered the power into their own hands in the Imperial Court. Meng T'ien stood in their way, they prevailed on the ruler to execute their will.

The slight military worth of the Great Wall was soon to be made evident. While the generals were deciding the quarrel over the succession to the Ch'in dynasty with the sword, disaster was brewing in the steppes of Turkistan and Mongolia. Mao-tun had united the scattered Mongol tribes under his leadership with cunning, deceit, violence, and shrewd maneuvering. He pointed to western China, weakened by inner struggles, as a common goal. Measureless booty waited there for him who dauntlessly stretched out a hand for it. And they did precisely that in 200 B.C. The flood of nomads riding shaggy, sinewy nags in formation surged against the border fortresses. The supposedly impregnable gates fell not only through a coup d'etat or betrayal. For decades no funds had been allotted for the maintenance of the Great Wall. Wind and weather had made many a breach by which the Huns now broke through. No help came to the garrison from the fortified camps, the troops were otherwise engaged, securing the throne definitively for the Han dynasty.

"The barbarians are coming!" The alarm was flashed across the land to the emperor's court. The emperor, knowing full well that the power of the invader could not be broken with armed might, offered to pay tribute. Mao-tun received thousands of rolls of silk, countless camels laden with silver and Chinese princesses, and turned them over to his followers. He received whatever he demanded until finally all China lay open to his reach. But what appeal did the parceled-out fields, the irrigation ditch and dikes, the working mass of industrious peasants have for the nomad? His horse could not race along there, his sheep and cattle could find no pasture. The mild climate would enervate him, the unaccustomed nourishment draw the strength from his veins. Far from the steppes the nomad was doomed to destruction. Mao-tun was shrewd, a worthy predecessor of the greatest of all Mongol leaders, Ghenghis Khan. He contented himself with the tribute, led the army back, and shifted the assault direction of his unruly troops toward the West—and thereby set off a chain reaction. The Huns exerted pressure on their western neighbors, who in turn were forced to seize new lands and set other peoples in motion. Thus the thrust propagated itself and after some delays reached Europe. Hence it was not the Wall that shifted the assault of the Huns from the south to the west, as is all too frequently asserted. Viewed militarily the Wall completely failed to accomplish its purpose. It was the tribute of the Han to Mao-tun that set the great migration of peoples in motion.

It was a clever chess move on the part of the Han. Relieved of the danger of the Huns, they could now concern themselves with the consolidation of their empire. China flourished especially under the long reign of Emperor Wu Ti. He ascended the throne in 140 B.C. In Europe Rome had just underscored her hegemony with the final victory over Carthage. Wu Ti had the Great Wall improved and he inserted a new section, over 300 miles long, in the West. Now the Wall stretched along the Tarim basin up to and beyond the Tunhuang oasis. The traffic along the Silk Road, later to become so famous, had just begun. The caravans made their way safely up to that point under the protection of the watch towers.

Wu Ti belonged to the category of enlightened rulers who learn from the past. He knew what an unreliable defense the Wall would be if danger should once more threaten from the steppes. He defended China on the perimeter. His armies penetrated deeply into the steppes and destroyed any organized formation of the nomads. There was no danger from the Mongols for as long as he ruled. In the far east he partially subjugated Korea. In the south he occupied the territories which we now call North Vietnam and northern Laos. He did not improve the capacities of the Great Wall for defense; his superbly trained armies were better suited for this purpose. The Wall was the symbol of China: China extended as far as the Wall, human beings lived within it, only barbarians without.

These barbarians did not necessarily have to be treated in an unfriendly way. On the contrary, directly beyond the Wall the Han dynasty promoted the well-being of the semi-civilized barbarians. They were to form a buffer against the uncivilized barbarians. As will be shown, however, this buffer turned out to be a double-edged sword.

When the mandate from heaven was removed from the Han (220 A.D.), the kingdom broke up into the states of Wei, Shu, and Wu. The Wall formed the northern border in the kingdom of the Wei, which was so busy with struggles for hegemony that it completely neglected the walls. The enemy stood in the south. The northern wall dwellers were supposed to serve as a buffer. But instead, unnoticed, they gradually filtered through the bulwark so that when the Wei finally had achieved mastery in China, in 280 A.D. the nomads within the Wall were already organized in a small kingdom.

What was the use of the new improvement of the Wall undertaken by Emperor Wu Ti? The enemy sat within the country's gates and refused to leave. Indeed, the emperor had to purchase domestic peace with them by paying tribute. The gigantic fortress installation was worthless. But China always had—and still has—an ally stronger than walls and armies. Sooner or later every invader was absorbed by Chinese civilization—that is, sinified. This also happened to those able-bodied nomads. By the time they established their power over all northern China in the fifth century, they were already so sinified that they built more than 700 miles of new walls against their own kinsmen who were still building nomad circular fortifications. But they were not decadent, as is attested by the victories of their horsemen over the peoples of the steppes beyond the Wall.

The news that first made Europe aware of the existence of the Great Wall dates from the fourth century. The Roman historian Ammianus Marcellinus reported: "The Chinese, makers of coveted silk, live securely under the protection of an all-encompassing wall."

Decades came and went. The blazing heat of summer and the icy cold of winter in Turkestan worked on the Wall. Besides, the walls had been erected with inferior building material. Many sections were merely reed fascines pasted together with clay. The walls increasingly gave way, the watch towers crumbled, the garrisons fell apart. But nobody bothered about this section so far away—even knowledge of its existence was lost. The central and eastern sections of the structure, built of stone, put up a stiffer resistance to the ravages of time.

East of the Red Sea, meanwhile, there arose a founder of religion whose followers were soon decisively to change the political map of southern Europe and north Africa: Mohammed from Mecca. In contemporary China a man arose to absolute power, Yang Chein, of whom the chroniclers will report: "His deeds shortened the rule of the dynasty to two generations but thousands of generations still will flourish because of them." Yang Chien (Sui dynasty) unified China for the third time.

The pattern with which we are already familiar repeated itself: a mighty emperor repairs the Wall, adds new sections and at the same time commands an army that scatters the potential besiegers of this Wall in all directions. Once more the bulwark is not the expression of a military hedgehog position, but the symbol of power, the demarcation of a sovereign domain. That is clearly discernible in the region of the gigantic arc with which the Yellow River tightly encloses the semi-steppes of the Ordos desert. It is a region which, if constantly irrigated, yields good harvests. Nevertheless, the peasant must fight constantly against the nearby desert. Shifting dunes must be made fast, irrigation ditches and canals must constantly be protected against silt. And if there is an extended period of drought, all this effort will have been in vain anyway. China's rulers repeatedly tried to retain possession of the whole loop of the river. All colonization projects fell through, although there was no lack either of volunteers from the densely settled regions or of financial support. Yang Chien finally drew the necessary conclusions. A new wall cut the loop in two and thereby handed over the Ordos steppes to the nomads.

The Emperor engaged 30,000 workers to build this section of the Wall. Nevertheless, and despite all efforts, the walls, after progressing a short distance, repeatedly came crumbling down. Were the gods displeased? The Emperor questioned the oracle: "Let ten thousand [wan] be buried in the foundation," came the reply. Thereupon he ordered that 10,000 workers be cemented into the wall. Are the chroniclers trying to fool us? Not exactly, although Yang Chien did have a bad press in educated circles. Actually, the emperor ordered a worker named Wan (ten thousand) to be thrown among the filling material. The death of this unfortunate worker changed the luck of the building contractors.

Yang Chien left the stage of world history in 618 with an assist from his enemies—and the northern defense installations were promptly neglected. They sank into such insignificance that the Arab traveler Abu Dulaf Ibn Muhalil did not find them worthy of mention. Yet he must have seen them because he traveled along the Silk Road to China. This was in the tenth century. China was again split up into different kingdoms warring against each other. The Khitans, a people of Tartar origin, established themselves in the north. And now something strange happened. Up to this point the Chinese had built bulwarks against nomads; now the Khitans turned the tables. Newcomers from the steppe peoples, and always concerned with preserving their own identity, they of all people built a branch wall to the Liao River against the Chinese. But they soon became aware of their own real strength and pushed southwards through their own wall.

Thc Sung dynasty finally decided to take the edge off the main weapon of the nomads, their mobile cavalry. The Sung ruled in the southern part of China and had been fighting with the Khitans since the eleventh century for their bare existence. They built a wall–granted, it was not an original idea. But this time it was a barrier of a special kind. They planted 3,000,000 fast growing trees along their northern borders up to the sea. Canals, dikes, dams, small swamps, and fish ponds, together with the meadows and rows of elm trees, formed an obstacle over 300 miles long. This barrier was supposed to stop the serried cavalry formations. But the Sung did not seem to be strongly convinced of their new defensive fortification because silk, silver, and daughters of the ruling house each year were loaded on caravans which set out on the road to the Khitans. They were by no means tributes. That would have signified losing face. They were presents. But no presents were sent back to the Sung in return.

Nevertheless, the Sung had constantly to "rectify" their front; even the wall of meadows could change nothing. But if the hare (Sung) fled before the fox (Khitans), the fox found himself fleeing from the wolf. The Juchens, booty-hungry and unspoiled Tungusic nomads, attacked the Khitans from the rear. But even the Golden Dynasty, as the Juchens called themselves, were not happy with their booty. In the steppes the tiger was already sharpening his teeth in order to devour them altogether. Temujin had not yet unified the peoples of Mongolia, he was not yet called Genghis Khan. How did the Chin (Golden) dynasty use the respite granted by history? They built additional walls against the threat of an attack from the North.

They especially fortified their capital, Peking. The triple belt of wall rose to a height of over 60 feet, a river supplied water to the moats of the city with its millions of inhabitants. Rectangular forts covered the four corners, each of which extended over a square mile and was connected with the city by a tunnel. It was a remarkable defensive installation. The section of the West Wall north of Peking received new rock revetments, the towers new tiers, and the gates facing north were cast in iron. The garrisons teemed with troops, weapons, and provisions. Wild horsemen take these bastions by storm? The very idea was laughable. And, in fact, for the first time the Wall fulfilled a military purpose. The otherwise irresistible cavalry formation of the great Khan camped impotently in front of the Dragon Gate that covered the entrance to Peking.

The "*Jagan Kerme*" (Mongolian for "white wall") held. Genghis Khan was still a novice in the art of laying a siege, but there were other possibilities, of course. A son galloped at full speed like a whirlwind at the head of a thousand horsemen to the neighboring gate. Sixty miles meant nothing to a Mongol horseman. If his horse lagged, he threw himself onto one of the horses being led. Sixty miles was just a little extra ride. Not so for the defenders of the Wall. Before reinforcements could arrive, the heads of the tower guards lay in the dust. With the Mongols inside the walls, in the rear of the Dragon Gate, those who could withdrew to Peking. The walls of the capital could not be conquered from the outside, not even by him whose motto, inscribed on a jade seal, read: "In heaven: God. On earth: the Great Khan, the might of God."

The siege lasted four months. Neither side could win. Then the emperor of the Chin recalled China's old and well-known remedy against the nomads: silk, silver, and girls. Richly laden, the army of the Great Khan withdrew toward Karakokum. A generation later Peking fell by betrayal to the nephew of Genghis Khan, Kublai Khan. He completed the conquest of China and assumed the Chinese imperial title in 1280, as Shih Tsu.

Under the Mongols, designated as the Yüan dynasty in

Chinese history, the Great Wall fell into a state of total neglect. What was one to do with it anyway? The greatest empire that had ever existed on earth extended from the River Vistula to Korea and from the Siberian taiga down to the primitive forests of Burma. Everywhere the Wall lay deep in the interior of the country. The Mongol Tumen were a far more effective defense than the longest and strongest wall in the world.

Marco Polo traveled to the court of the Khan. He was commissioned by Kublai to inspect the Chinese provinces and, as is known, he wrote a famous book about his experiences, but he makes no mention of the Wall. Why not? He must have passed by it but considered it of no importance. Perhaps, however, it belongs to the half of his experiences that he did not put down in writing. When at the hour of his death he was urged to recant all the lies and exaggerations that he had included in the description of his journeys, he replied: "I have told barely half of what I have seen." This is probably true: the Venetian wasted no words, for example, on the tea drinking of the Chinese–in medieval Europe tea was unknown.

The famous Arab traveler Ibn Batuta briefly mentions the existence of the Wall. But he certainly had not seen it, because in the fourteenth century one traveled more safely in Asia on the sea. The "Pax Mongolica" no longer existed, the individual parts disintegrated and formed a new independent kingdom. A Chinese dynasty, the Ming, seized power in the Middle Kingdom. Foreign dominion was at an end.

While toward the end of the fourteenth century the West opened itself up wide, paved the way for the Renaissance and therewith laid the foundations for its present-day dominance in the world, the China of the arch-conservative Mings encapsulated itself and kept the outside world at a distance. They were great times for the Chinese Wall. Construction crews renovated the crumbling walls and bastions along the whole length of the rampart. New offshoots arose, and the utterly neglected forts beyond the Wall again acquired importance. Together 15,000 of these isolated bastions formed a defensive chain in the

A huge caravan before the Chinese Wall: envoys of the Czar to the Emperor of China in the year 1692.

perimeter. Nearly all of the Wall that today's traveler sees was built by the Ming. Occasionally he will find on the revetment memorial tablets such as the one that lists the names of 130 officials and engineers who "worked together to erect this section of the Wall third class, 591 feet and 6 inches long. It begins north of the watch tower no. 55 (letter "Wu" of the Black Series). The autumn guard reported the completion of the wall section on the sixteenth day in the ninth month of the fourth year of Emperor Wan Li."

For the observer it is of no importance whether the Ming or the Ch'in built the area visited. Neither the mode of construc-

tion nor the material had changed. And if the Wall were broadened today, the same construction methods used by Meng Thien 2000 years ago would still be used.

For the Ming the Wall not only had a symbolic value, but was actually supposed to be a military defense position. Thus they not only tried to make the bulwarks and fortresses impregnable, but they also solved the problem of troops. What would be the use of the highest and strongest wall if there were no defenders behind the battlements, if the watchtowers were not manned? Fortified camps arose at close intervals directly behind the Wall. The soldiers were really peasants. They performed military duties only for a short period and otherwise tilled their own fields with their families. Thereby the burden of the costs of maintaining the army was notably reduced for the government.

The Ming rulers knew very well why they should attach an overriding importance to the Wall. Emperor Wan Li especially distinguished himself. He ruled for 38 years and built so much that some consider him the actual creator of the Great Wall. Wan Li had learned from history how wave after wave of booty-hungry nomads had poured in from the steppes. This rhythm would be repeated. This time the Chinese wanted to be prepared for the assaults with the Wall which in the meantime had grown to a length of some 3665 miles. Indeed, they were superbly prepared, and now had cannons at their disposal as well.

For almost three centuries China lived thus, content with herself and despising the world of the barbarians. But she was not left undisturbed. In the middle of the seventeenth century several European powers arrived on the Chinese coast. It was only reluctantly that the emperor agreed to the establishment of trading posts. The foreigners ever pressed for added concessions, and the Imperial Court hesitated. But the power of decision was taken away from it: in the north the Mongols had united themselves again under the name Manchu and were making preparations to take to the field for booty. Their resolve was strengthened by a strange omen: the lost seal of Genghis Khan appeared surprisingly at the assembly which was to elect the new Great Khan. A shepherd had found it in the steppes. It could mean only one thing: "Mongols, you will reconquer the empire of your forefathers!"

First it was a matter of overcoming the Great Wall. Emperor Li Tzu-ch'eng could view the attack with utter calm. After all, the old fire-eater general Wu San-kuei was defending the ramparts north of Peking. Wu was not only an excellent army commander but he also had an excellent knowledge of the Manchus. He had often supplied them with auxiliary troops to quell rebellious impoverished peasants in the interior. The emperor's throne was wobbly, a long-mismanaged economy had utterly ruined the country and forced countless people into banditry. There was always a rebellion flaring somewhere, but as long as General Wu was on the government's side, a successful attempt at revolution was inconceivable. While Wu and his army withstood the assault of the Manchus in the mountains north of the capital, Tsen, "whose beauty was like that of the full moon," waited for him in the homeland. She was his betrothed and the wedding was to take place at the end of the expedition. What did Emperor Li Tzu-cheng do? He added her to his extensive harem, and thereby sealed the fate of his rule.

The Manchus could not take the Great Wall by storm. Yet a demand to surrender now found a receptive ear with General Wu. The Great Wall had for once fulfilled its military purpose, and then this had to happen! It is not known whether the Manchus ever erected a monument to the beautiful maiden with the inscription: "Dedicated in gratitude . . ." The new dynasty ruled until 1911. Meanwhile, in Europe up to the end of the sixteenth century legends still prevailed concerning the Wall. In farthest Asia there was a Wall said to have been built by Alexander, who kept the peoples of Gog and Magog in check. Woe to the world if the bulwark should crack open one day! At the same time, there was a European who had seen the Wall with his own eyes and who had also written an account of it—the Portuguese adventurer Fernão Mendes Pinto, who in 1540 had shipwrecked on the Chinese coast. With great difficulty he made his way almost as far as Canton, where the authorities took him and, after charging him with vagrancy, sentenced him to a year's hard labor. He escaped from the labor camp in 1541 and published his experiences in Europe. He was considered an even bigger liar than Marco Polo, and yet there was certainly something truthful in his account. In 1615 the Jesuit Matteo Ricci sent back reports on the Great Wall, thus dispelling the legends. At the same time, an envoy of the Czar, a Cossack officer named Ivan Petlim, saw the Wall from Turkistan and provided information at first hand. He himself saw only a small section of the Great Wall, but Chinese traders informed him about the enormous extension of its construction. It was only at the beginning of the eighteenth century that a detailed map of the course of the Great Wall arrived in the West.

In 1907 the Wall was again being talked about in the West. On his Asia expedition Sir Aurel Stein had discovered part of the long-forgotten walls and towers that the Han had built 2000 years before in Turkistan. He found not only the wall itself, but also eloquent witness to the life of that time in the buried watchtowers and silted camps. The dry climate had excellently preserved everything, the oldest preserved paper, the remnants of silk, and reports of the troops. Among other things he also found a broken arrow in a tiny case with the accompanying letter: "Enclosed 1 (one) arrow, broken during practice shooting. Back to the depot in exchange for 1 (one) arrow, new. Signed–countersigned." Obviously the requisitions officer did not make it easy for soldiers even at that time. But the archeologist was far more impressed by another discovery.

"The obliquely falling rays of the setting sun uncovered a sign of the remote past. The line of the Wall was clearly set off for miles even where it was nothing more than an accumulation of earth. In these moments the eye became confused by a straight, furrow-like line parallel to the Wall and about 30 feet distant. Upon closer examination it turned out to be a narrow yet well-defined path that had been made in the earth by the patrolling guards through the centuries."

In the age of modern artillery the importance of the Wall dropped to zero. However, it played a central role in the Boxer Rebellion of 1900. At this time China was like a powder keg. People had had enough of tutelage, blackmail, and exploitation by foreigners, the white devils. All that was needed was a single spark to set off a huge explosion. The Wall provided it. Some American journalists, having little news to report at the time, concocted the story that American engineers were on their way to China in order to help with the demolition of the Great Wall. In some way the story, which had been long forgotten in the United States meanwhile, reached China. "The Americans want to tear down the Wall," screamed the headlines in the Chinese press, and the powder keg exploded.

The outcome is known. Acting in concert, the Western powers quelled the uprising and China was forced to pay the fantastic war indemnity of $320,000,000. It is an irony of history that this reparation later contributed decisively to the development of atomic missile weaponry in China. The Americans set up a student exchange with their share of the booty. From this fund came the stipend that financed Hsue-Shen Tsien's journey to the United States as a student in 1935. In 1955 he returned to his homeland as a missile specialist, albeit not wholly voluntarily. He was a victim of McCarthyism in America.

The Wall exercised a more direct influence on the group that entered history under the name of "Forgotten People." This came about as follows: about 50 miles west of Peking the mountains form a lozenge-shaped valley 15 miles long and 10 miles wide. It is a wild desolate region accessible only from one side. In 1644 it became the refuge and hideout of the remnants of the last Ming army fleeing the Manchus. Its existence did not remain hidden for long, but the victorious Manchus were magnanimous. They permitted the refugees to settle in the valley but they also strictly forbade them ever to leave it. They were serious about this prohibition: a new section of the Wall closed off the only exit from the valley. The elders of the village were allowed to visit the nearest city only once a year in order to deliver the tribute there. Up to the time of the Chinese Republic several thousand people lived in this enormous prison and preserved the customs and folkways of the Ming period.

If Emperor Shih Huang Ti once more saddled his magic steed and inspected his work today, after 2200 years, he would see much that has been changed but also much that has remained the same. As was his ideal, a central authority reigns, the rule of the great families and of the feudal nobility has finally been broken. Thanks to his restored and extended Great Canal, the peasants sow and reap in dry and rainy years. His

Wall does lie in ruins; only a few sections were renewed. Yet his aim of separating China from the outside world has been achieved. An invisible barrier closes China off from the outside world more effectively than the most perfect wall ever could. It is not for nothing that Mao Tse tung has been compared with Emperor Shih Huang Ti. I believe Shih Huang Ti would be gratified by it.

5.

The Oldest Seismograph: Dragon Heads and Toad Mouths

WANG HO laid the Book of Changes aside. Were his ears deceiving him? It seemed that the bell in the room, in which the earthquake weathercock had been installed, had pealed. No, it could not be. This bell rang only during an earthquake and he would have distinctly felt that. Wang Ho was the night watchman in the Imperial Chancellery for Astronomical and Calendrical Science.

In the capital city of Sian, chroniclers recorded the year that today we designate as 138 A.D. Wang Ho came from an influential well-to-do family which was even distantly related to the Imperial dynasty. His teachers had been very satisfied with him when he had successfully passed the important State examination in the capital. His poem in rhythmic prose had so greatly pleased the strict examiners that they gave him a mark denoting excellence. Therefore the highest State offices were open to him. But this meant that he first had to make good in lower positions. He had worked seven long years under the supervision of experienced masters; the years at the Imperial University had not always been easy. He had spent many nights painting characters on bamboo strips, which he then cut out with a knife. In the morning tiny heaps of bamboo chips attested to his industriousness. He had also copied, over and over again, the writings of Confucius and those of his commentators. How lucky that paper existed so that he could so thoroughly demonstrate his skill in calligraphy. Although hardly a generation had passed since the Chinese had learned to make these yellow sheets out of rags and plant fibers, already there were paper mills everywhere.

Wang Ho was satisfied with his lot. It was only a question of time until he would be advanced to more responsible positions. Yet he had to be careful not to jeopardize this promising career by frivolous behavior. "It is better to make sure that everything is in order," he decided.

The earthquake room was in the center of the building, its entrance sealed. Wang Ho hesitated a moment, then he tore the silk strip with the red seal of the Chancellery. Timorously he examined the huge copper cauldron. The foot of the vessel stood directly upon the foundations of the building and was therefore hidden under the floor. The moonlight barely illumined the room, so that it was difficult to recognize the dragon heads. The dragons, visible from the door, still held their individual balls in their mouths–nothing out of order here. But perhaps there was something not in order on the other side? The young scholar walked slowly around the richly ornamented vessel. Suddenly, his heart stood still; one of the dragon mouths was closed, and the ball that belonged to it now lay in the wide-open mouth of the toad below. An unheard of event! Up to now the earthquake weathercock had reliably recorded every quake and had also given the direction whence it had come. But now the earth had not shook, yet the dragon looking toward the northwest had dropped his ball. What could that mean?

Wang Ho felt panic beginning to set in. This had happened during his tour of duty, it could affect his career. Only the future could decide whether this would be in a good or bad sense. Right now Wang Ho had more important things to do than to think about that: the Imperial Court Astronomer had to be informed immediately about this incredible happening.

Had the brilliant invention of the Imperial Astronomer Chang Heng (78-139 A.D.) failed?

Only today, in the twentieth century, can we measure the importance of the scientist Chang Heng, this remarkable star in the heaven of Urania. In the year after his appointment as court astronomer he constructed a celestial globe, which performed the same function as a planetarium does today. It provided knowledge on the movements of the stars. In contemporary commentaries we read that this artificial heaven wondrously accorded with reality. Water power drove the armillary sphere around a polar axis, while a hidden mechanism allowed for different rotation velocities.

At the same time Chang Heng published cosmological ideas that are not outdated today. He represented the cosmos as completely empty except for the heavenly bodies. Crystalline spheres, as posited in the Middle Ages in the West, had no place in his world view. He explained solar and lunar eclipses as well as the phases of the moon: "The sun is like fire and the moon like water. Fire emits light, water reflects it. Hence the moon's brightness is produced by the sun's radiance; it is dark where the sun's rays do not strike it. The light emitted by the sun does not always reach the moon, since the earth can step in between–this is then called a lunar eclipse. When the same occurs with a planet, we call it an occultation. When the moon moves through the solar rays, there is a solar eclipse." Obviously Chang Heng recognized the correctness of the heliocentric world view.

This extraordinary man proved himself to be a master in all fields, whether he was active as poet, painter, philosopher, astronomer, mathematician, geographer, or statesman. As an engineer he was 1500 years before his time. Besides masterpieces such as the seismograph, he is also supposed to have constructed a flying machine with which he actually lifted himself into the air. (At least, that is what his contemporaries report.) Nor should it be forgotten that he, a sesqui-millennium before Cyrano de Bergerac, wrote a novel which today we would call science fiction.

The uncertainty lasted only a few days; then messengers arrived at the capital and reported on a serious earthquake a thousand li (around 400 miles) northwest, in Kansu. Thereupon Chang Heng's fame grew to unbounded dimensions. Not only did his seismograph accurately register every earthquake in the environs of the capital, but it even registered those in very remote districts of which no one in Sian had the slightest inkling!

In centrally administrated China this was of outstanding importance. Individual provinces were plagued by earth tremors and help had to be rushed to these disaster areas as swiftly as possible. Sufficient provisions for emergencies were at hand in government warehouses, since the State bought surplus harvests at fixed prices. Well-built roads ran from the capital in all directions of the compass and heavy cargoes could be shipped by way of an extensive canal system. There were no transportation problems in the well-organized State of the Han. But it often took too long until the messengers arrived in the capital and could report the disasters that had struck in the distant parts of the vast country. Thanks to Chang Heng's invention, the authorities could dispatch an expedition in the indicated direction even before the messengers had passed through the gate of the capital with news of the earthquake. Is it any wonder that simple people believed in magic? Indeed, it was also rumored that Chang Heng with the help of the copper kettle could not only establish the direction but also the position of the center of the earthquake.

All earthquakes were now recorded in the Imperial Annals ever since the horrendous earthquake in 780 B.C. had changed the course of three rivers. A total of 908 had been recorded up to the seventeenth century. The lists are a source of information for modern seismology. Thus, for example, the 32-year periodicity for the occurrence of serious earthquakes could be deduced on the basis of these records.

As with all peoples, superstition regarding natural events of this kind that penetrated so deeply into the life of human beings was rife in China too. Serious earthquakes were supposed to announce the end of a dynasty. As late as the year 1128, the seige of Sian, initially so promising, was broken off because of an earthquake. It was considered an evil omen—but obviously not for the beleaguered.

Although the astrologers tried very hard to predict earthquakes from the celestial constellations, no success attended their efforts. Moreover, it is a problem that not even the seismologists of the space-travel age have been able to solve up to now.

Chang Heng's invention—we have already made its acquaintance in connection with the determination of *pi*—was far ahead of its time. Even the most outstanding scholars found it difficult not to believe in supernatural powers when this first seismograph in mankind's history went into operation. Also unusual was the fact that the instrument was constantly in readiness. Otherwise, the general practice was to set up the desired experiment, determine the magnitudes to be investigated and that was it. Here it was quite different: the ball from the dragon head fell into the toad's mouth whether or not an observer was present. The earthquake weathercock was constantly on the alert. In our day we take it as a matter of course that instruments exist which function without a human presence, but a notion of this kind was still alien in the Middle Ages.

It is a stroke of good fortune that a rather exact contemporary description of the seismograph has been preserved: "A copper kettle with a diameter of six feet hid and protected the mechanism inside. Eight dragon heads were placed at regular intervals around the outside of the kettle. Each dragon held a metal ball between its movable jaws. Eight toads, cast in bronze, on the floor opened their mouths in the direction of the dragons. If an earthquake struck the locality of the kettle, the dragon that was looking in the direction of the oncoming wave opened its mouth. The ball fell into the toad's mouth, a bell rang, and the dragon's jaws snapped shut."

It seems that seismographs existed in Persia in the twelfth century. But we have no detailed information about them. Perhaps they were copies of the instrument in Sian.

Chang Heng's invention, dating back to 200 A.D., apparently was a towering individual achievement in China. After the overthrow of the Han dynasty (220 A.D.) everything went topsy-turvy. The three successor States, Wei, Wu, and Shu were busy fighting each other and neglected the sciences. Scholars, of course, could read about the weathercock in the records of the Han dynasty that had been preserved, but the mechanism and its operation remained unclear to the investigators of the time. Indeed, in the sixth century it required a genius to understand how it functioned. We learn more about this scholar in the history of the Northern Ch'i dynasty: "Hsintu Fang showed even in early youth an outstanding mathematical ability; everyone in his locality was full of praise for him. He was a man of great genius and often he was so immersed in reflection that he forgot to eat and drink, and even fell into pot holes when he went for walks. He told friends that when he was concentrating on mathematical problems, he 'did not even hear thunder.' "

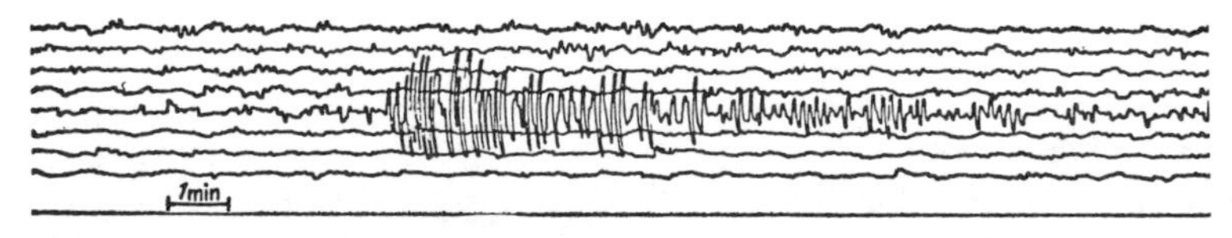

Sketch of a modern seismograph, reading from left to right; the stronger the earthquake, the greater the swerve away from the median line. This seismograph is shown registering an earthquake in California on December 25, 1951.

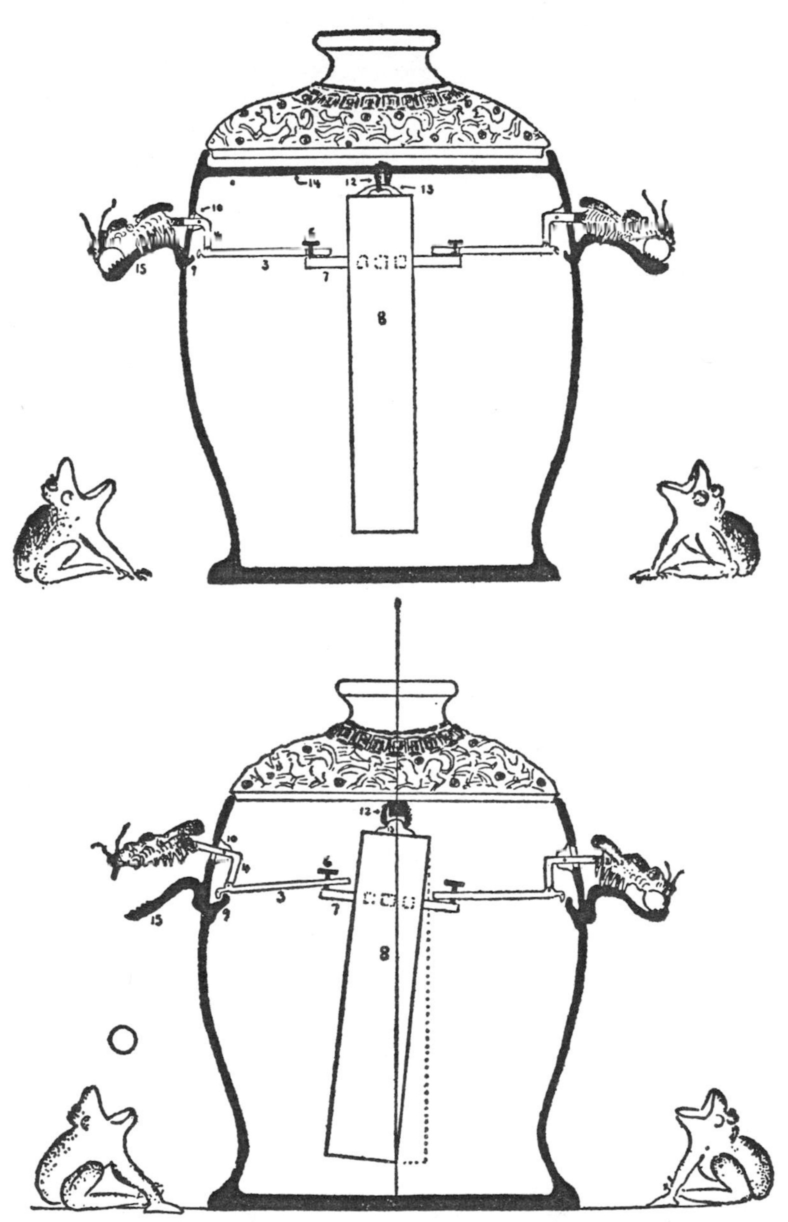

Cross-section of the kettle of the earthquake weathercock, showing the pendulum hanging from the domed cover and the lever mechanism that opened and closed the dragon's mouth. When the dragon's mouth opened, a ball dropped into the mouth of the frog below.

By the thirteenth century Chinese scholars entertained only vague ideas about seismographs. Thus Chou Mi concluded his account of this mysterious device with the words, "I cannot at all figure out how it functions and I would like very much to know who could enlighten me about it."

6.

The Compass

LONG Fan had his worries. For more than a week the army had been wriggling its way like an enormous worm across the boundless plateaus in a northward direction and still the thick blanket of clouds made it absolutely impossible to take bearings based on the position of the sun. Long Fan's job was to see to it that the army did not miss the pass across the Nan Shan mountains and that it avoided excessive delays resulting from detours.

Of course, the "south-pointing carriage" was at his disposal. This two-wheeled vehicle from the Imperial depots, built by the mathematician Tsu Ch'ung-Chih, had been assigned to the punitive expedition against the Tanguts, a Tibetan tribe.

A checkered history is connected with this carriage. Long Fan's thoughts wandered: Already at the time when the Chou still ruled over a united China—that was about 2000 years ago—there was talk of the vehicle whose wooden passenger always pointed south. No, it had nothing to do with magic; it was the work of a brilliant engineer. An ingeniously contrived system of toothed wheels and driving belts joined the movement of the wheels in such a way that the outstretched arm always pointed in the same direction, regardless of whether the carriage had changed its direction in the meanwhile.

Fiction and fact blend in the old writings on the south-pointing carriage. Indeed, for many hundreds of years scholars even doubted that it had ever existed, or they confused it with the other south pointer, the compass. Eight hundred years before, the Emperor had commissioned the court engineer Ma Chun to construct a copy of the legendary vehicle. Ma Chun was eminently suitable for this task, with no peer. He had already invented the hodometer as an aid for geometers: a drum chariot with mechanical puppet-musicians who beat the drum each time a li was traversed.

The reconstruction was successful to everybody's satisfaction, and later served Tsu Ch'ung-Chih, the great mathematician, as a model. Now the carriage was more than 500 years old. Its bronze gears still meshed flawlessly, the wooden, iron-sheathed wheels were perfect. Yet after long journeys the mute passenger was pointing too inaccurately. At times a wheel skidded on a curve. The more than 50,000 turns of the wheels since the last bearing based on the sun had been taken caused notable magnetic declinations no matter how carefully the artisans had given the same measure to both wheels.

The south-pointing needle, the compass, remained as an alternative. The steel needle in the tent, suspended from silken threads changed daily, pointed immoveably in a northwest direction. But did it really do so? The needle had to be rubbed repeatedly with the lodestone and always took a longer time before it came to rest after the impetus. The precious stone, moreover, broke into pieces one day, after a servant dropped it. The fragments could not transmit to the needle the mysterious property of pointing toward the south. Not only did the needle's force fade, but the magnetic declination–the natural deviation of the needle from the astronomical north-south direction–changed notably from place to place. Probably other valences applied here than those that obtained back home in the Imperial City. If only the sun would show itself!

Long Fan awoke from his brooding. Yang, the first servant, was standing before him. "The Marshal invites you to his tent, the subordinate leaders are already assembled. Your advice will be requested."

"Follow me, then."

In the Marshal's purple tent he was extended the greeting that was owed by protocol to a member of the Supreme Council of Philosophers. Then the Marshal told him: "Master Long Fan, I have decided to divide the army in two. General Tsai-yü will march with the 2nd and 3rd corps at a distance of 100 li parallel to us. How will he be able to hold on to the direction if the sun continues to hide behind clouds? We do not have a second reliable compass."

"Early tomorrow morning the fish will show you the direction." So saying Long Fan bowed deeply and left the tent.

"Yang, come here quickly! Start a charcoal fire, blow on it until it glows white, and then get the small crucible ready! We won't be getting any sleep tonight. Also get the Compendium of Military Technology!"

A little later the master was holding the book by Tseng Kung-liang which had appeared the year before. It did not take long before his eye fell on the passage he was searching for. "When troops run into murky weather or dark nights and can no longer hold the direction they should be led by an old mare." Long Fan smiled; even in such a modern work one found residues of superstitious times long gone by. He read farther, concentratedly: "Troopers can also use the south-pointing carriage or the south-pointing fish in order to find the direction again. . . Much is known about the method with the fish: a thin leaf of iron is cut into the shape of a fish. It should be two inches long and a half inch wide. Head and tail are pointed. This fish is to be heated in a charcoal fire until it is thoroughly red hot. Then it is to be picked up by the head with iron pincers and placed so that its tail points due north. In this position the tail is immersed in water several inches deep and cooled off. The cooled fish is to be kept in a tightly closed box. In order to show the direction the fish must be carefully placed in a basin filled with water so that it floats, whereupon its head will point southwards."

Satisfied, Long Fan set aside the Compendium of Military Technology. His memory had not deceived him. General Tsai-yü would find his direction.

Long Fan knew that the fish had to be cambered into the shape of a boat in order for it to float. The author of the book also knew it. To this day important war secrets are not completely entrusted to printed books.

Was the advice given by Tseng Kung-liang in 1044 A.D. in the Wu Ching Tsung Yao (Compendium of Military Technology) nonsense, superstition, or magic? Let us leaf through a modern manual of physics, published in 1969 under the title *Thermoremanence.*

"Thermoremanence is the magnetization that ensues when a substance is heated above the Curie point and then cooled off in a magnetic field. The magnetization direction of the thermoremanence corresponds to the magnetic field during the cooling period." What is described above in an up-to-date scientific formulation is the same effect that was also known to Tseng Kung-liang. If we add that the Curie point for the iron lies at 774° C, it will be clear that some individuals in China more than 900 years ago knew how to make a compass by utilization of the natural terrestrial magnetic field.

We may note further that in recent times the thermoremanence of volcanic minerals has been studied for the determination of geological age. This is connected with the fact that the earth's magnetic poles move in the course of centuries and millennia. Therefore, the compass needle does not point exactly in the north-south direction most of the time. This deflection is called the magnetic declination. We shall discuss this later. If the lava flowing from a volcanic eruption contains magnetic components, it will turn in the direction of the magnetic pole and thus harden into rock formation. If we know the course of the motion of the magnetic poles from other indices, we can determine the age of the lava from the direction of the magnetic components. This is a technique of geologic study that is being increasingly utilized by geophysicists.

The walls and the street plans of Chinese cities are a very special case of these so-called "magnetic dating" methods. When they were originally built, they were exactly oriented along the magnetic north-south axis with the help of the compass. Thus the direction of the walls can betray their age.

What about the south-pointing carriage? For a long time it was confused with the compass or even rudely relegated to the realm of fable. Nevertheless, the intensive studies of sources conducted by Professor Joseph Needham attest that as early as 1000 B.C. two-wheeled carriages existed in which sat a wooden figure whose outstretched arm always pointed in the same direction.

The knowledge of the mechanism that set the figure in motion was repeatedly lost, so that in several periods even Chinese sources relegate this vehicle to the realm of magic. Nevertheless, such a carriage was reconstructed from time to time, the last time probably in 478 A.D. A detailed description of the machinery involved is also extant, but presumably the author was not the engineer himself because the mechanism could by no means have functioned in the way it is described.

Obviously it is no problem to build a carriage of this kind in the twentieth century. In 1947 the English engineer George Lanchester designed a reliable functioning model according to the Chinese pattern. However much Mr. Lanchester tried to come up with new designs, it would have been a faulty construction without a differential gear. Does that mean that this essential component of the modern automobile was already known in ancient China? We do not know.

This aid for adhering to a prescribed direction could hardly have achieved much importance. On long journeys it becomes unreliable. Even trifling differences in the circumferences of both wheels, tiny inaccuracies in the gears, the skidding instead of the rolling of the wheels, all these contingencies add up to notable shortcomings. The arm would no longer point in the originally oriented direction. Yet this principle has not been forgotten. It suffices to read the description of a modern tank, for example, that of the "Leopard." In combat its cannon constantly points to the originally aligned target regardless of the tangled course taken by the tank during the combat operation: here again is the south-pointing carriage of the ancient Chinese.

In ancient China when a man built a house, a prefect set up a village, or an emperor founded a city, a part of the costs flowed into the pockets of the geomancers. We know little about this guild which was already well-established under the Han dynasty, more than two thousand years ago. Its books were burned in the seventeenth and eighteenth centuries by the Jesuits as works of deception and devilry.

Their job was to indicate lucky sites for human habitation. Neither site, nor the direction of the habitation, nor the dimensions were established without their considered counsel. In order to understand why this was so, we must make a brief excursion into Chinese thought. In traditional Chinese philosophy, the whole universe is subject to the interplay of the two complementary primal forces Yin and Yang. They complement each other as male and female, light and darkness, inhaling and exhaling, slow and fast, condensation and evaporation. One is not conceivable without the other. This dualism does not entail a value judgement; it is not the polarity, say, between good and evil. All creatures of the animal kingdom are born under the rule of one or the other primal force, they belong either to Yin or to Yang. Only man assumes a special position, in him both rule at the same time.

The doctrine was presumably created by Tsou Yen (c. 400 B.C.). He taught in the Chi-Hsia Academy on the Shantung peninsula at the same time Plato and Zeno were teaching in Athens. There are indications that Tsou Yen took over the basic ideas of his school from Persia. In addition, he also developed the theory of the five elements which we shall discuss later.

The followers of Yin and Yang, who are sometimes called naturalists, soon merged with the Taoists, while the disciples of

Confucius took over the hypothesis of the five elements. In contrast to the Taoists, who mostly led self-absorbed, meditative lives, the naturalists stood with both feet in practical life. As geomancers they offered their services to anyone, from peasants up to the emperor. Their knowledge was to serve the purpose of helping man to live in perfect harmony with Yin and Yang on earth. The ideal of inner and outer peace, eminently worth striving for, was obtainable only "when Yin and Yang exactly take the place due to them, when calm and peace reign. If the scale tips to one side, the wind arises. If they collide, the thunder roars; if their paths cross, lightening flares; if they are confounded with each other, fog and clouds are begotten."

To be sure the geomancers also busied themselves with magic, prophecy, and the interpretation of dreams, and the choice of a site was definitely accompanied by hocus pocus (which, of course, raised the prestige of the guild). Yet we must give them credit for having been the first successful urban planners. The houses, hamlets, and towns of China are harmoniously integrated with their surroundings. Man and nature form an aesthetic unity. They were clever people, who in all their considerations never forgot the well-being of the inhabitants: the proximity of water, gentle breezes, charming vistas, circulating air, and so forth.

It was not only the art of landscape configuration that enabled the guild of geomancers to survive more than 2000 years; the flexibility of their views also played an essential role in this respect. They viewed the world as being in a state of constant change, and they were convinced that harmony between Yin and Yang at different times could be achieved through different methods.

If both forces did not successfully satisfy each other and peacefully co-exist, misfortune was bound to follow. When the engineer and general Meng T'ien built the Great Wall on orders of Emperor Shih Huang Ti, he could not always bring its course into harmony with Yin and Yang. Military considerations had to be taken into account; the foundations reached deeply into the earth and thereby violated Yin. He traced his unfortunate end—the successor of Shih Huang Ti sent him the Silken Cord—to the disharmony that had thus been engendered.

The modern seeress avails herself of a crystal ball or a tea cup, astrologers make their readings from the constellations of the planets, futurologists interpret computer results. The geomancers also had an aid: the Shih, the bipartite diviner's (or magic) board. The apparatus rested upon a square earth plate, made of wood and bronze, whose surface was painted with magic symbols as well as with constellations and compass points. It was surrounded by a rotatable round heaven plate. Its outer edge was decorated with the 24 constellations, while the seven main stars of the Great Bear were represented in the center of the heaven plate.

The arrangement accorded with the orthodox world view. The world, washed on all sides by the ocean, had a square shape and was subject to the primal force of Yin. Above it arched the celestial globe, where Yang ruled.

When the heaven plate rotated on its central pivot, it came to a stop after a while and the tail stars of the Great Bear pointed to one of the signs of the earth plate. In this form the diviner's board functioned not much differently from the roulette in a modern gambling casino. This changed when the heaven plate—presumably in the Han period (202 B.C.-220 A.D.)—was replaced by a crude spoon. It was no misguided change. One can also signify the stars of the Great Bear as a spoon seen in profile, and in English, of course, this constellation is called the "Great Dipper." Thus we see that the heaven plate of the magic board was replaced by its essential symbol, precisely the pointing constellation. If the center of the earth plate and the underside of the spoon are polished, the spoon can rotate freely. The handle (or tail of the Great Bear) serves as a pointer.

We do not know the name of the geomancer who replaced the heaven plate by the rotating spoon. Nor do we know the

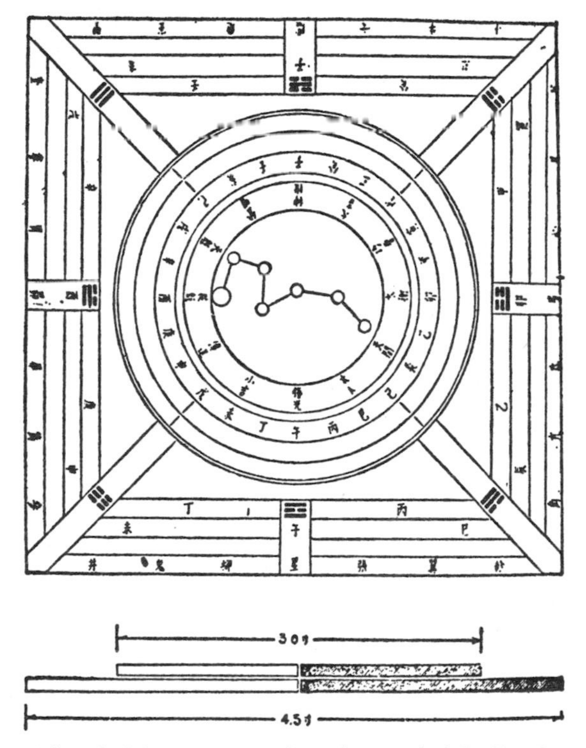

The diviner's board of the geomancers from the period of the Han Dynasties. The round plate of heaven, constructed on a central pivot, revolves around the rectangular plate that represents the earth.

identity of the person who introduced the second and the most highly important change of the "Shih": he did not fashion the spoon out of wood or bronze but out of a mineral, magnetite. If the support (or ground plate) and the spoon are mirror-smooth, the spoon comes to a stop (after rotating) in the direction of the terrestrial magnetic field, in a north-south direction. This was something quite different from the accidental position on the old magic board. With the new apparatus a geomancer could really impress the uninitiated.

As the name indicates, the iron ore magnetite, also called lodestone (Fe_3O_4), is by nature magnetic, that is, it attracts iron. There were written accounts from China concerning this phenomenon as early as 300 B.C. Thus it was not at all a misguided notion to carve a spoon out of the magnetic substance that attracts iron.

Our unknown geomancer was probably also surprised to see the handle always pointing in the same direction. What an excellent aid for making prophecies and telling fortunes! That it was used for purposes of this kind is confirmed by the complaint of the interregnum Emperor Wang Hang (9-23 A.D.): "Heaven has given me the throne as shown by the handle on the board. How then can the army of the Han defeat me?" He had ordered a geomancer to question the magic board as to whether heaven would now remove the mandate to rule. But the negative report received did not prevent the soldiers of the Han dynasty from taking the palace by storm. The Emperor died by the sword and with him the attempt to change the social structure of the country, which would be delayed for nineteen centuries.

The magnetic spoon was rather unserviceable as a directive aid on journeys. Even with the most highly polished surface the friction was still so great that the spoon, when still, could decline considerably from the north-south direction. To reduce friction, the spoon was made to float upon a piece of wood in a basin filled with water. From this arrangement it was only one small step to the wooden fish with the lodestone in his belly—the first liquid compass. A needle along the extended middle line of the fish led to greater accuracy in the determination of direction than the head of the fish had been.

At least since the first century it was known in China that needles could also be changed into magnets by contact with magnetite. Indeed, ultimately the quality and thus the price of a scrap of lodestone was determined by measuring its power of attraction with needles. If it held ten needles, one hanging to the other, it was worth its weight in gold.

At the time of the Sung dynasty (960-1234 A.D.) the strength of a lodestone was determined with still greater precision. The left tray of a scale for weighing gold was made of iron. The scrap to be tested was placed upon it and it weighed down the scale. Then the right tray was loaded until it dropped. This is a method that is still used today by students of physics to measure a magnet's power of attraction. The chief difference is that today students do not determine the power of attraction of a mineral, but of an electromagnet.

Needles soon lost the properties transmitted to them which made them magnets. It was not until the fifth century, when the highly valued steel needles were imported from India, that the

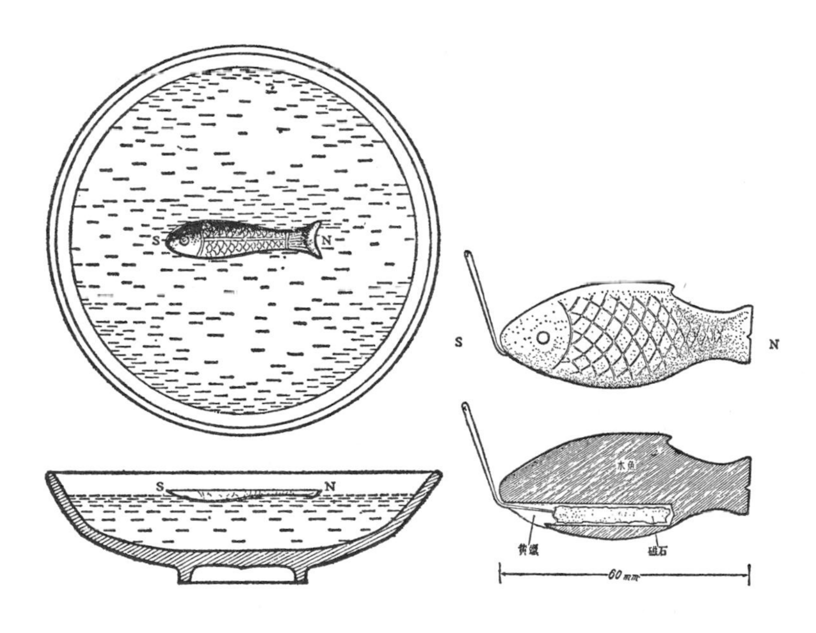

Early form of a magnetic compass. The wooden fish carries a lodestone in its belly. When it floats the projecting needle points south.

swimming fish could be replaced by the considerably more accurate magnetic needle. These Indian needles were so light that they swam in water if they were first smeared with fat and carefully placed on the water's surface. To be sure, it took a while until the magnetic needles oriented themselves in the direction of the terrestrial magnetic field, but after that they held to their position very exactly. Perhaps this is the origin of the child's game in which thin needles are placed on water following which the future is foretold from the play of the shadow on the bottom.

Another method of mounting magnetic needles was to suspend them from an individual cocoon thread of the silk worm: thus the dry compass was invented. But the slightest disturbance sufficed to make these freely rotating needles vibrate around their position of rest, whether they were suspended from a silk thread or rotated around a point on a broader spot. This was a great disadvantage aboard ship, so that Chinese mariners through the seventeenth century preferred the liquid compass which quickly controlled the oscillations and vibrations of the dry-pivoted magnets. In Europe and in Arabic cultures the dry suspension found a use in navigation from the outset. And today the dry compass has been almost completely replaced by liquid compasses.

"Magicians rub the point of a needle with lodestone, whereupon it points to south. But it always inclines slightly to the east, and does not point directly south. The needle can also be made to float on water, but such a position is rather precarious. It can be balanced on the finger nail or on the rim of a cup where it can turn effortlessly. But since these supports are hard and smooth, it easily falls off. It is best to suspend it in the middle of a single cocoon fiber of raw silk. It can be attached with a tiny piece of wax no larger than a mustard seed. If the needle is in a windless spot, it will unfailingly point south. Among the needles there are some that point north after being rubbed. I have both kinds of needles. No one can explain the principles of these things." This citation comes from the Encyclopedia of Shen-Kua and was written in 1088 A.D. It is an absolutely reliable historical source, and we shall be referring to it more often later, especially in connection with the art of printing. There are still older references to the compass in Chinese literature, but this is the oldest clearly interpretable passage—a good hundred years before the compass is mentioned outside the Middle Kingdom.

On the other hand, there is information about the magnets themselves dating back to prehistoric times and preserved in

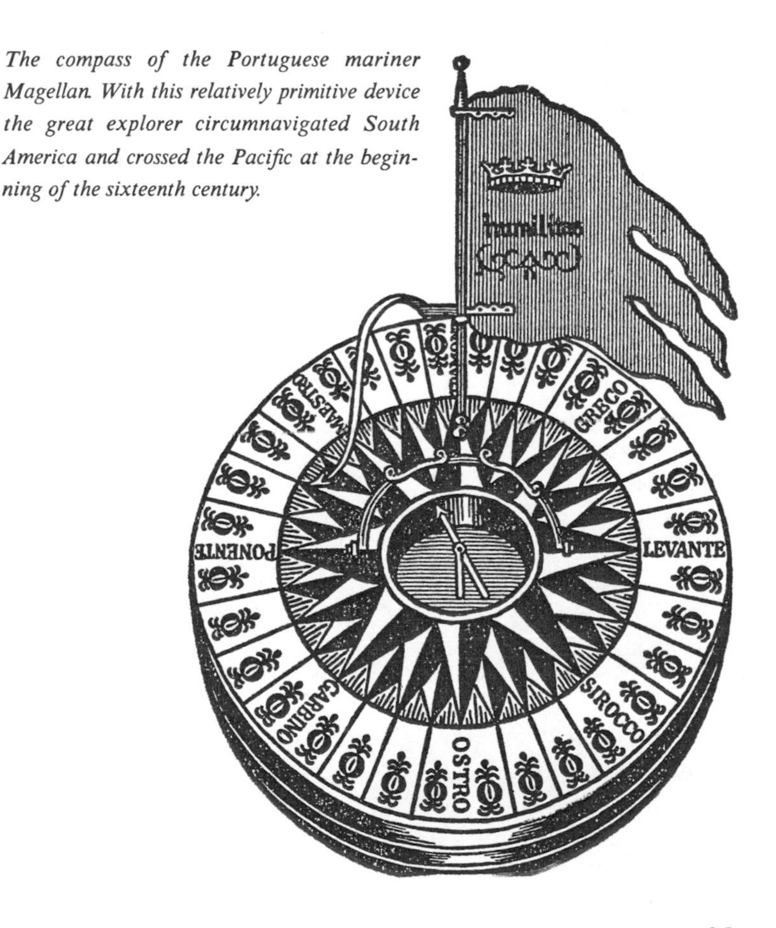

The compass of the Portuguese mariner Magellan. With this relatively primitive device the great explorer circumnavigated South America and crossed the Pacific at the beginning of the sixteenth century.

myths. Examples are the legend of the magnet mountain in whose proximity ships sank in the waves because the mountain tore the iron nails out of the planks, and the fairy tale of the king who first made his visitors pass through a door made of lodestone in order to detect whether they might be carrying concealed weapons.

The first concrete reference goes back to the sixth century B.C. Thales of Miletus qualitatively described the properties of the magnet. Three hundred years later a Chinese source transmitted a similar description. But here the quantitative measure of the strength of magnets is already given, indeed even how many needles linked one to the other the magnet can hold.

The modern oculist would certainly be very surprised to know what "modern" methods his colleagues in Kaifeng used over a thousand years ago. They extracted metal fragments from eyes with the help of lodestones. Children who had

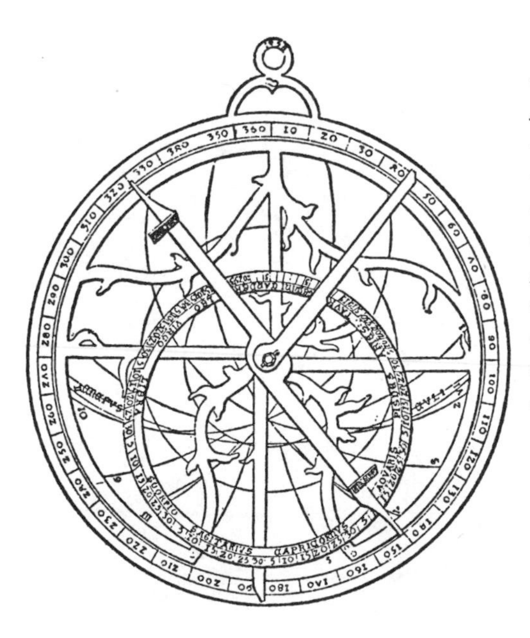

The astrobium of Johann Regiomontanus from the year 1468. Many captains of that time—including Columbus—could not make use of this complicated precursor of our present-day sextant. That was the task of a mathematician with knowledge of astronomy.

swallowed iron needles were made to swallow small magnets in the hope that both alien bodies would expel each other. But they could not have had much success inasmuch as the magnets were too weak.

As we have already seen the directive property possessed by the suitably shaped lodestone was already known in China at the time of the birth of Christ. In fact it was so well known that the constancy of the south-pointing indicator became a metaphor used in love poetry.

Outside China it was not until the year 1190 that a description of the compass needle was transmitted, by Alexander Neckan. In 1269 Petrus Peregrinus of Maricourt (Peter the Traveler) published a detailed description of the compass, the clarity of which was not to be surpassed for decades. In fourteenth-century Italy the form of the compass card (*rosa ventorum* or wind rose), still serviceable today, was developed. One or more compass needles were attached firmly to the card. In the sixteenth century this improvement of the compass reached the Middle Kingdom and was immediately and fully adopted.

Hence the Chinese can undoubtedly claim the glory of discovering the compass. But let us read the aforementioned citation once more. It not only describes the north-south directive property of the compass needle, but Shen Kua also knew about the declination of the needle, that it systematically declined from the astronomical meridian. That is highly interesting since this discovery was attributed to Columbus, which would date it 400 years later.

We have good reason to assume that the discoverer of the New World had no inkling whatsoever of the declination of the magnetic needle. This is all the more amazing because in 1450 the artisans in Nuremberg who made the portable sundials took account of the magnetic declination. The compass that was built into each one received additional marking, deviating slightly from the north-point. The sundial had to be adjusted accordingly if it was to give the correct time.

There are still other sources that show that the declination of the magnetic needle was known to the Chinese long before other peoples. Thus, for example, in 880 A.D. the magic board of the geomancers, adorned with concentric circles, was broadened with an additional circle on which the four chief points of the compass were painted. The added signs are the same as those already existent, but they are all shifted around 7.5 degrees to the east. Thus they compensate for the effect caused by the fact that the instrument does not point in exactly the north-south direction.

At the end of the ninth century in China the declination amounted to 7.5 degrees. In the twelfth century another painted circle appeared, this time deviating around 7.5 degrees to the west from the original arrangement. Thus the declination had changed its direction.

How does this accord with modern research? The declination of the compass periodically changes in amount and direction. A full period, that is to say, the time elapsed after which the declination again exhibits the same amount and direction, lasts several hundred years. At the present time the declination in Central Europe diminishes yearly by 0.1 degree.

In the age of mercantilism Europe overcame the earlier time lost with great strides. Mercator (1546) predicted the existence of the magnetic pole. Coulomb (1785) discovered the laws of magnetostatics. Alexander von Humboldt (1829) introduced the first world-wide description of terrestrial magnetism (geomagnetism). Gauss (1832) invented the magnometer. Maxwell (1864) set forth a theory of magnetism that is still valid today. Amundsen in 1903 reached the north magnetic pole on Boothia Peninsula.

Our excursion into the history of the compass would be incomplete if we did not ask ourselves how this important navigational aid reached Europe from China. For there is no doubt that the road runs along this direction. The answer may seem to be the sea route from China by way of the Mohammedan world into the Mediterranean area.

That would have been possible. The following passage, part of a government regulation of the year 1086, proves that at this time the compass had rooted itself firmly in Chinese navigation. "The pilots of ships are familiar with the configuration of the coasts; at night they steer by the stars and at day by the sun. In murky weather they look at the south-pointing needle. In addition they use a hundred-foot line with a hook at the end, with which they bring up mud samples from the sea bottom. The consistency and smell help them to determine their present position."

Up to now there are no known references in Chinese literature indicating that the compass was used in navigation before the tenth century. On the other hand, it is difficult to imagine that the junks, which in the eighth century undertook voyages extending as far as the Persian Gulf, could have managed without this navigational aid. Perhaps there are even more surprises in store for us in this respect. On the other hand, we must consider that China was always an agrarian country and navigation on the high seas seldom played a great role. Only a fragment of the trade was plied along sea routes. The rivers and canals in the interior of China were considerably more important for the transport of goods. Another reason could be in the fact that steel needles that met the requirements of ship compasses could be manufactured only relatively later. Thus before the tenth century the use of the south-pointing needle was left overwhelmingly to the geomancers, land surveyors, and astronomers.

Chinese ships did not sail as far as European waters. Thus a direct contact and therewith the transmission of knowledge directly from the source could not take place. Arab merchants were the intermediaries–for hard cash, of course. Nevertheless, in no Arab or Persian writing is there mention of the compass before the year 1240, and at this time it had already been known in Europe for a half century. Everything points to the conclusion that the compass was not found among the spices, silk, precious stones, porcelains, and other valuables which the Arab *Dhaus* took over from the junks.

At all events, two hundred years, perhaps more, went by before the compass reached Europe. Why did the transmission

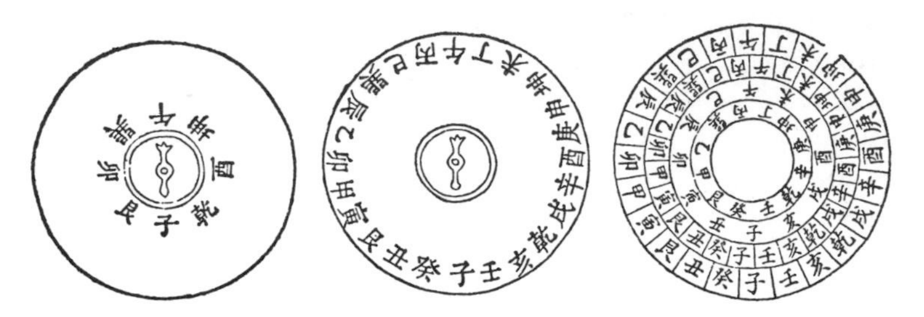

Left: two simple Chinese compasses with 8 and 14 directions; south is at the top. Right: geomancer's compass plate from the early tenth century. The inner circle shows the secondary cardinal points; the second ring indicates the main cardinal points. In the outer ring the symbols are combined. The characters on the third ring from the center are moved 7.5°, thus taking the easterly magnetic declination into account.

take so long? This is very astonishing when we consider that the importance of the compass for navigation can hardly be overestimated. An equally important invention, the cannon, spread with lightning speed in all directions.

Let us even consider both sides: the giver and the receiver. The Sung dynasty (960-1234) ruled at the turn of the millennium. Art, science, and technology were promoted by a superbly managed administration. Nevertheless, news of the progress made arrived only slowly to the outside world, since the Chinese Wall was closed. This was due not so much to the free will of the Middle Kingdom as it was to the pressure of external enemies. The few Chinese who left the country were almost exclusively seekers after truth who walked in the footsteps of Buddha and paid little attention to worldly inventions such as the compass. (There was one exception, of course, printing—but we shall discuss that later.) The south-pointing needle found itself primarily in the hands of the geomancers, who were not particularly keen about making it known to others. All these factors must be taken into consideration when we ask why the compass required more than 200 years for transmission as far as Europe.

But that is not all. Many thousands of miles farther west, in Europe, the compass would have been of little use at the turn of the millennium. European rulers were all too much involved with internal problems to feel the need to broaden their horizons beyond the West. The emperors of the Holy Roman Empire were in a state of constant struggle with the Pope and the rebellious feudal lords. Vikings threatened the coasts in the north, and Slav peoples pressed on the eastern borders. The bourgeoisie and cities able to defend themselves were just developing, and internal trade between Milan, Burgundy, Novgorod, Byzantium was in the process of being built up. Three hundred years were to go by before the creation of the Hanseatic League. Europe's gaze was still inner directed, and the Crusades, which had just been launched, did little to effect a change in this respect. Europe was not at all prepared for taking over the compass, and perhaps it would have made short shrift of it as black magic. The compass would have been welcome only to the Vikings. By way of Ireland they had pushed forward to Greenland, settled there, and finally discovered the North American continent. They sailed regularly to Labrador in order to cut down the wood that was so coveted in Greenland. Navigational aids for these journeys were less important than we might think. After all, on clear days from Greenland one can sight the mountain tops of Baffin Island across the Davis Straits.

All indications argue against the hypothesis that the compass reached the West in the twelfth century through Arab intermediaries. Did it perhaps come from the north? Had the Vikings invented it independently of the Chinese? Had not these bold sea-farers on their voyages oriented themselves by the "sky stone"? Doubtlessly they found their course in the often overcast North Atlantic, as proved by their voyages to Iceland, Greenland, and finally Labrador. Was the "sky stone," perhaps, the compass?

The question was answered several years ago. Compass and sky stone are fundamentally different. If we grind calcite

($CaCO_3$), a mineral also found in Iceland spar, from a particular angle and look through it, we can discern the polarization direction of a light source. The sunlight that pierces through clouds is polarized at different degrees, according to the angle formed by the sun, the clouds, and the observer. The most strongly polarized light is emitted from the spot in the cloud-covered sky that is at a right angle to the sun. Accordingly, the owner of a stone of this kind can determine where the sun is in the sky and thereby determine his course. In other words, the Vikings did not invent the compass.

No matter how we figure it, the compass did not reach Europe by the seemingly simplest way, by sea. Only one possibility remains: the land route. And this is not so cockeyed a notion as it might seem if we bear in mind that the compass in China was primarily used by magicians, geomancers, and astronomers. These people traveled frequently enough, over the most important traffic artery of Asia, the Silk Road. They brought an endless stream of wares from northwest China to Egypt and Asia Minor. So, after it passed perhaps into the hands of the astrologers and magicians of Kanchow, Turfan, Samarkand, Nishapur, Baghdad, and Palmyra, the compass reached the West. The proof? The din of the caravans has faded away, the caravansaries have crumbled into ruins, the freight lists of the merchants have been lost. The old Silk Road is no more. Yet there is one intriguing clue: up to the seventeenth century, astronomers used only compass needles whose tips pointed south—as in ancient China.

Thus the history of mankind's first pointer instrument ends with a question mark. And when twentieth-century man sits in the cockpit of a jet or in the control room of an atomic energy plant, does he ever consider that the ancestor of all these pointers is the handle of the Big Dipper we know so well?

7.

Kites, Rotors, Balloons, and Parachutes

THE autumn of the year 19 A.D. brought many difficulties to Emperor Wang Mang. He had just brought the potentates of his country to reason through cunning, persuasion, and an occasional application of violence, when the Turkoman horsemen again invaded the northwestern territories of his kingdom. The Chinese Army proved too ponderous and slow-moving to cope with the extraordinarily mobile enemy who overnight could change its position by more than 200 li. The time when the Han armies put plundering nomads to flight was no more. The shadow of the Hun leader Mao-Tun cropped up threateningly from the past.

Should he, Wang Mang, continue the work of the Ch'in and also broaden the Wall around the Middle Kingdom in the northwest and strengthen it by additional towers? The provinces of Chao, Yen, and Chin had been living in peace for more than two hundred years. The Great Wall—called the "White Wall" by the bellicose nomads—could not be forced by cavalry formations. The Turkomans, moreover, had no knowledge of wall-breaching machines. What they could not conquer in a wild sword-brandishing onslaught remained out of their reach.

Nevertheless, the troops manning the bulwark were more important than the bulwark itself. The past had repeatedly shown that the Great Wall by itself was of scant military value. In conjunction with well-trained troops, however, it could play an important role. If the enemy was routed on the perimeter, the armies could rely on the supplies in the warehouses and arsenals inside the Wall, and in emergencies they could even win a breathing spell behind the bastions.

But at the moment the kingdom's finances were in disarray, and the granaries empty as never before. In addition, Wang Mang had to take another factor into consideration: by many feudal lords, although not by the majority, he was viewed as an upstart emperor who, as former chancellor, had usurped power. His supporters reported to him that his overthrow was expected in the not-so-distant future. If he should dispatch a strong army to the northwestern borders now, that could be the germ-cell of a general uprising against him. Would the leaders of these garrisons, so far from the Emperor's surveillance, not make a deal with his domestic foes? The nobility were merely waiting for a suitable opportunity to dethrone him. If only he had destroyed the nobility root and branch, like the rulers of the Ch'in period! Instead he had hoped to build his rule on the support of the people that owed so much to him. His hopes had been dashed and, after a decade of rule, he could depend neither on the hierarchy nor on the people. What was to be done?

Wang Mang turned his thoughts to the world immediately around him; a servant was kneeling at the entrance. "Tu Lin awaits the command of the Ruler."

"Show him in!"

"He comes at the right time. Perhaps he, the Sage, knows a way out," Wang Mang said to himself.

Actually, the Council of Philosophers was at the Emperor's disposal at will, but he also had the privilege of calling for consultation individual persons whose counsel he trusted.

Indeed, Wang Mang had been making increasing use of this privilege.

The first advisor to the Emperor appeared, subjected himself to the customary ceremonial, after which they discussed the situation. Tu Lin's habit was to speak the unadorned truth and he did not hold back from making a pessimistic appraisal of the developments in the northwestern provinces. Nevertheless, he did see a possibility for improving the situation:

"The inventive spirit of your people is inexhaustible. Is it not far superior to that of the barbarians? Therefore, this time too, brain power should succeed in defeating sheer brawn and raw violence. Fifteen years have gone by since you last summoned the scholars of the kingdom to the capital. At that time the philosophers, astronomers, geomancers, and engineers displayed wondrous and amazing things. At that time you alone wanted to promote the sciences; now it is required by the crisis facing the kingdom. Why not summon them to the capital again and ask them this time about effective weapons that can be used against your external foes?"

"Tu Lin, you are the pillar of my kingdom. Your counsel is good. As a token of my thanks and also so that everybody may recognize my esteem for you, you may harness another pair of horses to your carriage. And dispatch messengers with my seal with orders that in eight weeks the scholars must demonstrate their inventions for fighting the Turkomans."

The Emperor was visibly satisfied. His advisor was right. Why all these worries when the sciences were flourishing in the Middle Kingdom?

Tu Lin had not yet come to the end of his visit. He opened a small case from which he took out a silken scroll: "I have not come empty-handed. The head archivist gave me this notation on the Chi-Kung people. Allow me to read it to you:

"The Chi-Kung are an ingenious people. They know many things that are hidden from other peoples. They travel through the air on huge aerial carriages with lightning speed. When the Emperor Tang—that was two thousand years ago, my Lord—ruled the world, a westerly wind bore the aerial carriages as far as Yuchow where they landed. Tang had the carriages dismantled and hidden in the depots. The people too easily believed in supernatural things, and the Emperor did not want to alarm his subjects. The visitors remained ten years and then reassembled their carriages, loaded them with the Emperor's gifts, and flew away on a strong easterly wind. They safely reached the country of Chi-Kung, 40,000 li beyond the Jade Gate. But nothing more is known about them.'"

"As a child I heard about the Chi-Kung," the Emperor said. "I took it to be a fairy tale. But perhaps the aerial carriages really exist. The world is full of wonders and no limit is set on the human mind. If our troops had such aerial carriages, they need have no fear of the Turkoman horsemen."

Tu Lin bowed: "Perhaps one of the scholars knows the secret of the Chi-Kung."

The yellow tent with the standard of the ruler towered above all others on the bank of the river. The scholars of the kingdom had obeyed the Imperial order. But it was not only a gathering of dignified scholars. Charlatans and illusionists hoped to snatch a good share of the expected rewards for themselves.

One of these, looking sublime in a long white mantle tied around his waist with a pilgrim's cord, was the first to be called. He brought knowledge from India and offered to lead an army across water without getting its feet wet.

The Imperial order was that he should first of all demonstrate this feat in person. After a long meditation the Indian confidently climbed the ladder of the landing pier which rose high above the river. With eyes closed, holding on firmly to his staff, the pilgrim stepped forward without the slightest hesitation–and fell smack into the water below amid the general laughter of the onlookers. The court officials were well educated and could clearly distinguish between assertion and knowledge. But nothing was to be left untried. The Indian who had taken an involuntary bath–fortunately the water at this spot was rather shallow–nevertheless received an honorary robe from the Emperor.

Meanwhile, the bodyguard marched up, and the Emperor sat

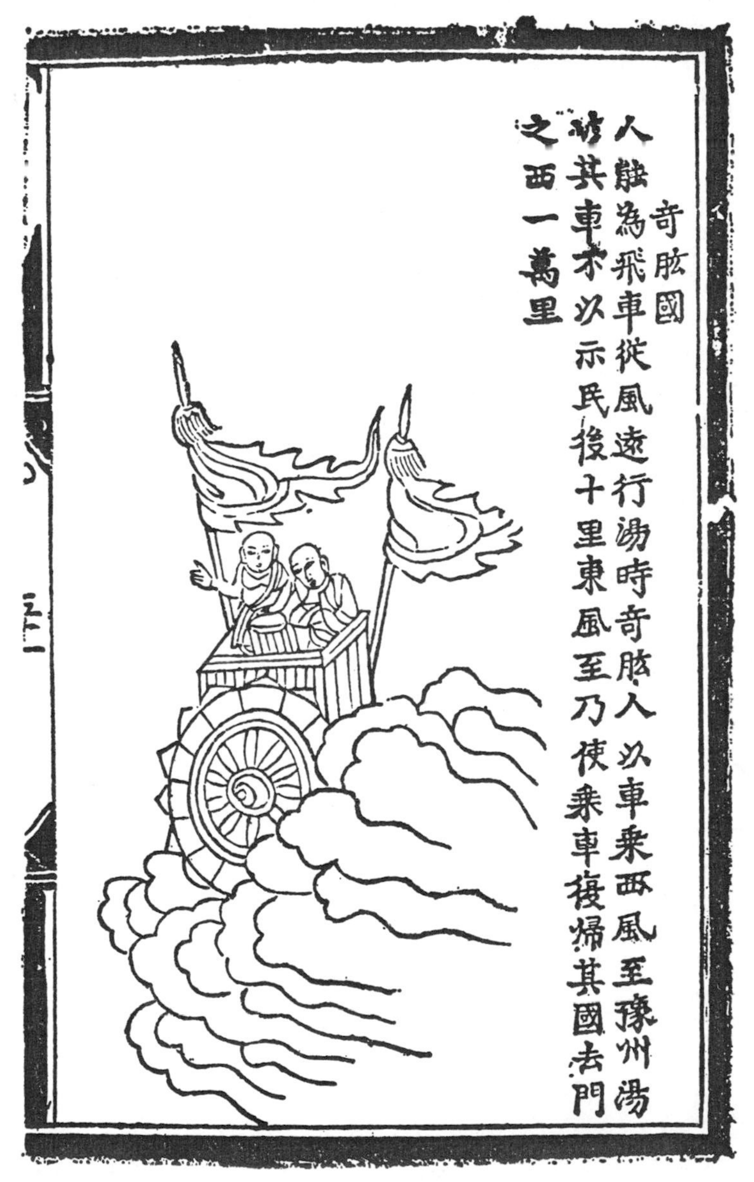

奇肱國
人能為飛車從風遠行湯時奇肱人以車乘西風至豫州湯
破其車不以示民後十里東風至乃使乘車復歸其國去門
之西一萬里

The oldest picture of the legendary Chi-Kung people and their aerial carriages. From the Chinese encyclopedia of the year 1430 A.D.

down on his elevated throne. He beckoned Tu Lin to his side. "Well, does anybody know the secret of the Chi-Kung?"

"No, we have found nobody who can build an aerial carriage. Nevertheless, someone is getting ready to try Mo Ti's invention over there on that wooden tower."

"Mo Ti?" repeated the Emperor in astonishment. "Do you mean the honored philosopher, the father of Mo Tshia? He lived four hundred years ago!"

Tu Lin took the manuscript from the hand of the waiting attendant, and unrolled it. "I have here an excerpt from Han Fei Tzu. Permit me to read it to you:

'Mo Ti built a wooden kite which took him three years. It actually could fly, but after a few days it was wrecked. His pupils said: how ingenious indeed is our Master that he can make a wooden kite fly! But he answered: it is more difficult to make a wooden ox-yoke peg. For this all you need is a piece of wood, eight-tenths of a foot in length. Further, all it requires is less than a day's work and nevertheless one can pull a load of two tons. The ox-yoke peg will last many years after covering wide stretches of land and pulling heavy loads. But I have worked three years on this kite which fell apart after one day. Hui Tzu heard about this conversation and said: Mo Ti no doubt is very ingenious, but perhaps he knows more about making ox-yoke pegs than about making kites."

The first Councilor commented: "Mo Ti himself reported that the engineer Kungshu Fan built a kite according to his instruction which remained aloft three days. A still larger kite, made of bamboo, even lifted a scout in the air. The observer was supposedly very useful during the siege of hostile cities."

Wang Mang nodded reflectively. Such a device would be useful indeed. To be sure, his enemies controlled no cities worth mentioning, but a man high in the air could reconnoiter the Turkomans from afar and spy on their assault formations.

Kites were not unknown in the Chinese army. When a battle had to be fought with the superstitious nomads, many kites were sent aloft, suspended from silk threads after which light musical instruments were sent up and attached to them. The rattlers, clappers, and flutes resounded eerily in the wind and took the steam out of the nomads. It was a guaranteed method to steer the enemy in a particular direction where the attack could best be intercepted and contained.

Meanwhile, the preparations at the tower were completed. The light-footed crew stood ready at the tow-rope. The silhouette of a powerful owl was etched darkly against the sky. The bamboo framework was covered over with silk, the observer was suspended between the wings. His body, solidly fastened to the framework, was barely half as long as the kite. The Emperor raised his hand.

"It flies! It flies!" roared the spectators. And, in fact, the crew ran against the wind for a while, at the same time releasing the tow-rope through their fingers. The kite with its human cargo rose higher and higher. Suddenly a cry arose from the crowd; the owl tipped over on a wing and plummeted into the river with the silk clattering in the wind.

Tu Lin shrugged his shoulders: "An accident, the gods were not well-disposed toward it. Earlier I saw it come down to earth safely. Without any observers, the kite would certainly remain aloft. Tomorrow we will demonstrate how tiny signal flags can be pulled aloft on it. Will you now deign to direct your attention to that place over there?"

Tu Lin turned his gaze to his left where the oddest-looking object was standing. It was man-sized but it brandished powerful bird wings. At all events that was how it looked, since feathers completely covered the underlying light bamboo framework. Feathers also covered the body, behind which dragged a tail-like contraption. Now the inventor who was raising both pinions almost perpendicularly, flew off the ground with a jerk, slipped his feet into the tail-frame while at the same time flapping the wings. This time the crowd was silent, so that Tu Lin's voice could be heard clearly: "He claims that with such a device he can fly more than a thousand li every day."

This turned out to be something of an exaggeration. After a few dozen wing flappings, the bird-man fell exhausted to the ground. Obviously, he had not wholly unraveled the mystery of bird-flight. The Emperor nevertheless rewarded him with considerable honorary offices and enrolled him in his permanent entourage. His enemies would be very impressed by this abortive demonstration, and perhaps someday he would really fly.

Although discouraged he stayed on to observe how tiny fires were lit under silk-spun lanterns. The colorful balls rose aloft and sailed on the light breeze that was blowing, but many plummeted to the earth in flames.

"Well-suited for signaling, but unfortunately they are no substitute for the aerial carriages of the Chi-Kung," said the Emperor as, resignedly, he gave the sign to break up the gathering.

Did Wang Mang sense that his star was sinking? In 9 A.D. he, the Chancellor, had seized the throne of the Han. He had not employed violent means in order to enjoy the exalted privilege of wearing the yellow robe of the ruler, but had merely persuaded the great of the empire that he was ideally suited for the job—as odd as that may sound to Western ears. Such a thing never occurred in the history of the West; indeed the idea of acquiring a throne except by way of inheritance or violence was absolutely inconceivable. Yet in China, this was possible. After the despotic rule of the Ch'in had been broken, the Han wanted to find the ideal mode of government. After all, it was the highest duty of the Emperor to make possible a virtuous, honorable life to all the subjects of his empire through his rule. Only under this condition would Heaven preserve his mandate to rule. The person who cooperated in the achievement of this goal won the highest fame and honor.

The Ch'in rulers believed that the people could live in virtue only through the strictest regulation from above, and the brutal suppression of non-conformists. Their fate indicated the practical unfeasibility of this view. The Han wanted to rule more through exemplary deeds and actions; they wanted, together with the official hierarchy, to offer the people a model worthy of emulation.

At the time of the birth of Christ this system, which found its highest perfection a thousand years later under the Sung, was still not fully developed. It was a time of experiments with the State form, and—so the reasoning went—why not give a successful Chancellor a chance to carry out plans that sounded so persuasive?

But no one had foreseen how far Wang Mang would go in his reform. He introduced astonishing changes. First of all he set free all male slaves, a decree that embittered the great of the Empire, who, of course, sabotaged it in any way they could. Then he broke the power of the large landowners, by distributing the giant holdings among the tenant farmers and the liberated slaves. The people, full of gratitude, worshipped him from the outset, but he just as surely earned the undying hatred of the proprietors. The stratum that today we call the Establishment constantly intrigued against him. Nor would it shrink from the idea of taking foreign mercenaries into its service in order to bring about his overthrow. But this was not so easy since Wang Mang had foreseen something of this kind and, as a countermeasure, called in all gold coins which the Imperial Exchequer exchanged for bronze ones (a measure that had repercussions even in far-away Rome).

The Chinese Wall was widely opened. A direct consequence thereof was that trade flourished. Heavily laden caravans moved slowly along the Silk Road, opened in the first century A.D. Silk was the most important commodity which by way of many middle-men ended up mainly in Rome. The beautiful ladies on the Tiber could not get enough of it. But how was one to pay for the endless stream of seductive silk? Rome's strength lay in its battle-tested legions, not in goods; it could offer the silk producers at the eastern end of the caravan routes little in the way of exchange merchandise. Today we would say there was a negative balance of trade. And even at that time recourse was had to the same expedients: Rome exported gold. Thus more and more of the yellow metal arrived in the Middle Kingdom—and Wang Mang hoarded it. The outflow of gold assumed such proportions that Tiberius (14 B.C. - 37 A.D.)

forbade Romans to wear silk clothing. Thus Rome's gold reserves were kept stabilized at least temporarily. Wang Mang's rule cast a long shadow.

Emperor Wang Mang stood alone. The officials proved to be corrupt, the Army leaders lent a willing ear to the conspirators. His decrees were increasingly distorted on the way from the capital to the provinces. To this were added external difficulties. Mountain rivers burst their dams, overflowing and devastating enormous expanses of territory. Droughts plagued other districts, epidemics destroyed livestock. These were plagues that had always swept over China at regular intervals. But this time, thanks to the effectiveness of the propaganda of the upper classes, everyone blamed the Emperor. Obviously Heaven had withdrawn the mandate from him. A revolt broke out and spread throughout the country with lightning speed. The few troops that had remained loyal to the Emperor could not control the situation, and in 23 A.D. Wang Mang met a violent death in his palace. He had believed in his lucky star to the last.

Man's age-old desire to fly weightlessly in the air like a bird and to escape the limitations of earthbound existence is alive among all peoples. In many Asiatic religions, the highest stage of self-absorption is reached when the spirit releases itself from earth's gravity in magic flight. In most myths wing-bearing human beings rise to heaven or are borne across the firmament with carriages harnessed with fabulous creatures. The witch riding through the air on a broomstick is another imaginative variant.

The inventor who hoisted himself several hundred yards above the ground before the eyes of Emperor Wang Mang believed that man could lift himself into the air like a bird. Such a possibility easily suggests itself to those who study nature and her organisms. Yet up to the present time there is still no apparatus with which avian wing-beating can be satisfactorily imitated by human muscle-power. Models have been constructed, and daredevils have hoisted themselves aloft with pinioned contrivances of this kind, but fundamentally the problem has not yet been solved.

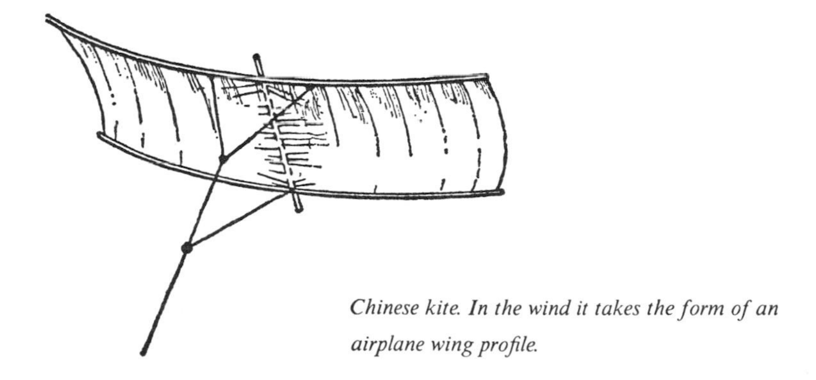

Chinese kite. In the wind it takes the form of an airplane wing profile.

The principle of the kite seemed promising from the outset. The harnessed framework of a kite can be raised aloft with the help of a long cord and a favorable wind. The three component forces, tow-rope, wind, and weight, can work together in such a way that a lift occurs and a flying device that is heavier than air rises. This is an ancestor of the modern airplane. The first model airplane, built by George Clay in 1804, had kite wings.

Although the philosopher Mo Ti is supposed to have invented the kite in 380 B.C., there are a number of references that are still older. But perhaps Mo Ti and his contemporary Kungshu Fan were the first to build a kite that could carry a man.

A clear account describing how a man hoisted himself aloft with the help of a kite stems from the year 549 A.D. The army of Wei was besieging the fortress city of T'ai. In order to get an idea of the defense installations inside the walls, an observer was sent aloft on a supra-dimensional kite. This was not exactly a pleasurable mission for the man in the breezy heights, because the besieged peppered him with arrows.

Another account, from the chronicles of the year 559 A.D.: "Ken Yank had Yuan Huang-Tou and other prisoners bound to paper kites in the form of an owl. They had to fly down to the ground from the Tower of the Golden Phoenix. Yuan

Huang-Tou was the only one who succeeded in flying as far as the Purple Road and in landing there safely." The distance between the Tower and the Purple Road is known: Yuan Huang-Tou covered 5 li or about 2 miles on his flight. The crew on the tow rope must have been exhausted.

Kites, frequently modeled after animals, are repeatedly mentioned in the Annals. Not only did they serve as favorite children's toys, but they also lent themselves to many uses. We have already mentioned their use as signals. Fishermen also used them to cast their bait far from their boats. And with their aid one could throw lines across a ravine and make it possible to build a bridge.

At the siege of Kaifeng (1232) the beleaguered had recourse to a means that is generally believed to be an invention of our time: they printed propaganda-leaflets, sent bundles of them aloft on kites, and released the leaflets over enemy lines. A contemporary observer wrote a sentence in this connection which twentieth-century commanders surely have not read: "If the generals hope that they can conquer the enemy with such methods, they will find that very difficult." The Mongols captured the city and measured the surviving prisoners on the axles of their carts, that is to say, they cut the head off of anyone whose body extended beyond the pin.

In Europe the kite appeared as a child's toy in the fifteenth century. It had slowly made its way to the West through the Islamic empire—according to the traditional records. But had it perhaps already arrived four centuries earlier, not as a toy but as a top-secret military device? According to some accounts kites were employed at the battle of Hastings (1066) to deliver messages.

In 1772 the kite entered the history of the natural sciences: Benjamin Franklin proved with its help that lightning is an electrical discharge between clouds and earth.

At the end of the nineteenth century an Australian built the box-kite, a construction with considerable lift-power. The rear chamber served as a steering mechanism and stabilized the flying device. After the turn of the century observers soared up to more than three thousand feet above the ground with a series of box-kites. The world record in kite-lifting was established on August 1, 1919, in Lindenberg (Frankfurt/Oder district). A train of eight box-kites achieved a height of some 28,000 feet. To be sure, this record could be broken today. Under favorable meterological conditions off-shoots of the jet stream roaring in the stratosphere reach heights of up to six miles. If a kite reaches these air currents, it can rise another six miles without difficulty–if the tow-rope holds.

The invention of the propeller-driven airplane limited the further development of the kite. Nevertheless, the kite also played a role, albeit limited, in World War II: German submarines sent aloft the most perfected kites, the gyro-kites, with an observer, thus extraordinarily enlarging their range of sight.

The kite was not the only toy to reach Europe in the fifteenth century. Another flying toy delighted children: the bamboo dragonfly. Four wings were set at right angles on a thin, wooden axle. A cord was wound around the wood which was then mounted in a catch-device after which the wound cord was pulled with a sudden jerk–the prefiguration of the helicopter then propelled itself aloft. This is a child's toy that can still be bought today in an up-to-date version, equipped with a plastic rotor and spring-propulsion.

The knowledge behind this flying contraption was also culled from a close study of nature. The seed of a sycamore tree is rotor-shaped and flies over 150 feet even in a light breeze.

Wind-power had been harnessed in the service of man in China since time immemorial. Windmills represented one of the principle power sources for machinery. But they bore no resemblance to the mills that are so familiar to us from pictures. Their vanes turn around a perpendicular axis.

Is it possible that a kind of helicopter can be developed with the help of this rotor?

It is really not that simple, of course, since something is still missing: the moving force. After all, no wind blows strongly

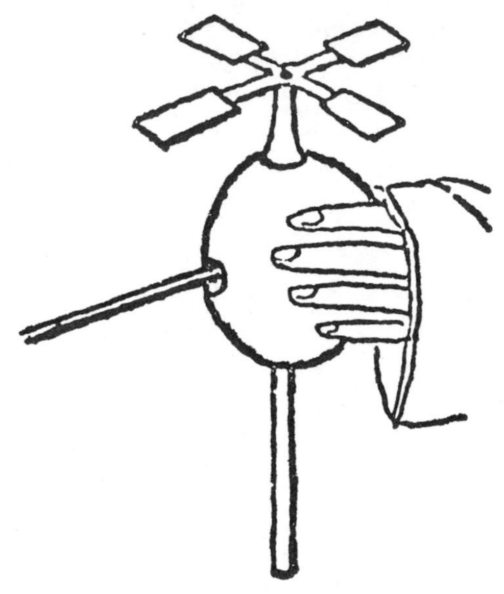

Medieval variant of the bamboo dragonfly. When the string is pulled the rotor spins up and away from the casing. (Section of a painting from 1460.)

and constantly enough to impart moving force to this helicopter. The Chinese did not possess a machine independent of spatial position. This machine had to be light and powerful. Because of its low efficiency and great weight, the tread-wheel did not come up for consideration.

Can a larger model be built according to the principle of the bamboo dragonfly? Professor Needham is of the opinion that a passage from the work of the scholar Ko Hung (c. 320 A.D.) can be interpreted in this sense. The Imperial astronomer and master of the mechanical arts Chang Heng (78-139 A.D.) himself has reported that he hoisted himself aloft in a contraption with rotating rotors and a built-in power mechanism.

This passage, therefore, should not be dismissed lightly, even though no details of any kind on this flying machine are to be found. No doubt Chang Heng was the most brilliant engineer produced by the Han dynasty. The seismograph that he built was extraordinarily ahead of his time: not only did its mechanism remain incomprehensible to contemporaries for hundreds of years, but it took more than half a millennium until something comparable was created. It is our loss that an exactly detailed description of this flying contraption has not been transmitted to posterity, as is the case with his earthquake weathercock.

What are the real facts about the aerial carriages of the Chi-Kung? Can anything more be said about them today? The Chinese chroniclers also added illustrations of the aerial cars to their accounts of the mythical people of Chi-Kung. Interestingly enough, the circumference of the car's wheels are not smooth but toothed. Do we see here the first propeller? The observer detects between the axes structures that could serve as gearings. It would unduly stretch the reader's imagination and patience to identify these pictures with actually existing aerial carriages. Moreover, these drawings were made at a much later time than the accounts that have come down to us. But the conception seemed already to have been clear to the ancient Chinese: rotors and a moving force were required for flight.

How did the development proceed in the West? In the fragments of Archytas of Tarentun (fourth century B.C.) there is talk of a flying model. Nothing is said about propellers and rotors. Perhaps it was the first ancestor of the model aircraft. Other interpreters believe it involved a jet propulsion with expanding vapor.

The rotor appeared in Europe only at the end of the Middle Ages, as a toy. But it was not long before the natural scientists fundamentally solved the mysteries of its flight. In 1784 Launoy and Bienvenu built a model with contra-rotating propellers, driven by a bow-drill device.

One hundred and fifty-three years went by before this development found its crowning achievement. Heinrich Focke and Georg Wulf presented to the world the FW61, the first serviceable helicopter. The Focke-Wulf works were far ahead

Borata
Chinarum gens admodum ingeni-
osa eße perhibetur, adeo ut currus
excogitarint fabricaverintque,
quos velis ventisque per campos
et loca plana uti navigia per
mare derigere optime norint

Drawing of an imaginary land sailing carriage by Mercator, 1613.

Another drawing from a work by Speed, 1626, shows several of the four-wheeled carriages sailing.

of their time. A sure sign of this is that several of the world records established by them were still unmatched in the 1950's.

It was the invention of Gottliebe Daimler and Carl Benz in 1883 that first allowed man to imitate birds, with propeller or rotor. But before that man tried another method, for which nature offered no model whatsoever to him: the balloon.

The first balloon filled with hot air, made by the Montgolfier brothers, rose in France in 1783. In the same year the invention took Charles and Robert over 3000 feet high—and brought them back safely. The Montgolfier balloon became increasingly safe, as attested by the fact that during the siege of Paris they regularly flew couriers and tons of mail over the Prussian lines. It was, of course, a one-way traffic, for the French capital could not be reached from the outside, since the wind determined the aerial route.

This dependence on the vagaries of the weather was eliminated first with the help of the steam engine, then with the gasoline motor. The time of propeller-driven airships began. Graf Zeppelin's dirigibles crossed the Atlantic. They provided paying passengers with far greater comfort than the Jumbo Jets

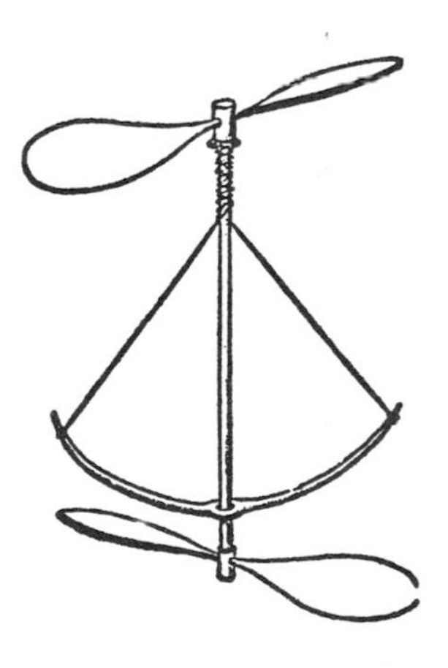

Simple helicopter model with two rotors. Launroy and Bienvenu, 1784.

can ever offer, completing the voyage from Germany to America in two days. The crash of the "Hindenberg" in 1937 at Lakehurst brought this development to a close.

But let us go back to the ancient Chinese. We recall that when Emperor Wang Mang in 19 A.D. convoked the inventors, there was also talk of flying lanterns. Are they perhaps the first beginnings of balloon travel? There is a Chinese text (200 B.C.) that describes how to make a body lighter than air: "Take an egg and empty the shell carefully, then ignite a tiny piece of wood in the hole in order to produce a strong air current. The egg will rise in the air by itself and fly away from the spot."

Thus the principle of the hot-air balloon (in miniature) was known. What was more obvious than to replace the eggshell by a covering of silk and later of paper? The flying lanterns served the people as amusement and the military as a signal system. The Mongols, who were simultaneously the conquerors and the pupils of the Chinese, employed these means of communication in 1241 when they inflicted an annihilating defeat on the combined German and Polish cavalry at Walstadt near Liegnitz.

Chinese sources make no mention of flight experiments with large baloons, but many Annals are still unopened and we should not exclude the possibility of future surprises. The knowledge, the building material, and the artisan skill at any rate were present in ancient China.

If today, in the second half of the twentieth century, Wang Mang again reviewed the accomplishments of inventors, almost all his wishes would be realized. Man still cannot walk on water, but amphibious vehicles offer an excellent substitute. Feeding an army is also a problem in our day (nourishment through pills is still far off), but dehydrated foods can be preserved for an almost unlimited time and do not weigh down the supply services. Bullet-proof vests also exist, but modern infantry weapons have long since nullified this mark of progress. Man requires neither kites nor bird feathers to hoist himself into the air. Moreover, Wang Mang could have his

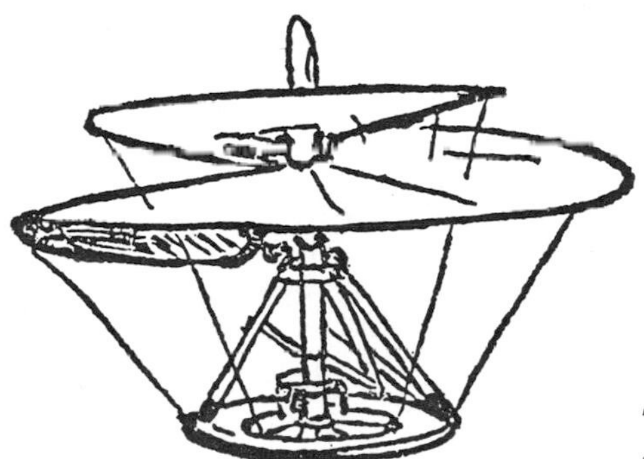

Sketch of a helicopter by Leonardo da Vinci.

choice of aerial carriages, from Jumbo Jets to freight helicopters.

But they could change his fate as little today as they did in his time. Heaven's mandate is still removed from the ruler who does not succeed in living in harmony with his people—at least if we take the long view.

8.

A Cheap Substitute: Paper

"I am sending you the works of the philosopher Hsu on ten scrolls. Unfortunately, I cannot afford a copy on silk. I make do with one written on paper." So reads the notice which the scholar Tsui Yüan included with the script scrolls for his friend.

His friend will have understood. Silk was expensive in China in 200 A.D.;the only alternative material for handwritten copies was wood or bamboo–and these were much too heavy. Besides, literate Chinese were slowly growing accustomed to the new, light substitute.

Slowly is perhaps not the right word, since it was only a few years before (105 A.D.) that the Court eunuch Ts'ai Lun had officially reported to the Emperor on the new writing material. The raw materials for its fabrication were not expensive: old fishnets, rags, tree bark, hemp, and grasses were available in huge amounts at little cost.

Nor did the manufacturing process present any excessive difficulties for the artisan: the basic substances were macerated into a watery pulp, the papermaker master added an agglutinate, and his apprentices, using finely meshed nets or flat pans, scooped out the pulp in thin layers. Once dried, the almost white sheets yielded an acceptable writing material for the characters of the calligraphers.

Naturally silk remained the better—and more expensive—material. The basic structure was finer, the details of the brushed characters were more clearly discernible, and the silk scroll could take more text than an equally thick writing material made of paper. The material fashioned from the cocoon of the silk worm was also considerably more resistant and durable than paper. Any child could tear a sheet of the substitute material. Silk, on the other hand, withstood the strongest grip; it was untearable. Nevertheless, paper possessed decisive advantages: it was cheap and could be produced in practically unlimited amounts.

The demand for writing material was enormous. The officials of the Empire alone required extraordinary amounts. The Emperor had made it a rule that all official events were to be recorded on paper so that even many years later one could reliably refer to any matter. The petitions and claims of the populace had to be presented in writing. For this purpose the State maintained special offices in which scribes wrote down the wishes of illiterate petitioners gratis.

The mass production of a cheap writing material was not only important for the apparatus of the officialdom developing under the Han dynasty. Science also profited from it. For the first time scholars could afford to have several copies of their work reproduced by scribes. The price of replicas would no longer rise to exorbitant heights because of the cost of exclusive silk. The wages paid to copyists scarcely entered into consideration as a factor of production. We do not exactly know whether Ts'ai Lun really was the inventor of paper. Perhaps he had merely given the Emperor an account of this new development. Otherwise how are we to explain that there were already references to the use of paper from the year 102 A.D., hence three years before the memorandum of the court official?

We owe this important invention to the inhabitants of the province of Szechwan in the valley of the mighty Yangtze

Obtaining paper sheets from bamboo. (Japanese drawing.)

Kiang. What would modern society be without a reliable, durable, and space-saving communications medium? Certainly, baked clay tablets, polished bamboo, and smoothed wooden boards were all suitable for receiving and preserving characters. Nevertheless, the libraries assembled from them took up lots of space, and thus every great philosophical work was quite "weighty." Silk was better: there was space for comprehensive notation on a small scroll, the storehouses of the libraries could be drastically reduced. Unfortunately, however, silk not only served as writing material for scrolls. The demand for silk clothing fabric was much greater. China delivered silk, by way of the caravan routes, as far as Rome and the constantly rising demand coupled with the long journeys and the numerous middlemen kept the price high. Naturally, the expensive material was always available for imperial ordinances and the works of important poets and chroniclers, but two things still had to be invented for the broad dissemination of the written word: paper and printing.

The papermaking craft spread with lightning speed over China. If at first it had been only a cheap substitute for silk, the situation soon changed. Manufacturing methods were improved, the basic substances carefully chosen and inks developed which fixed the characters permanently on the paper. Since the seventh century, papermaker masters impregnated the sheets with a beef extract. This left behind a weak yellow color tone but reliably prevented insects from devouring the paper. According to government regulation, all documents had to be treated with this insecticide.

Silk played almost no further role as writing material when printing began in the eighth century.

The early date of the invention of paper is not to be doubted. Modern measuring methods confirm the accounts of contemporary Chinese chroniclers. A paper fragment dug up in the Great Wall in March 1931 by the archeologist Folke Bergmann can be objectively ascribed to the second century A.D. with the help of the radio-carbon method of dating.

Paper found its way to the West, as did so many other

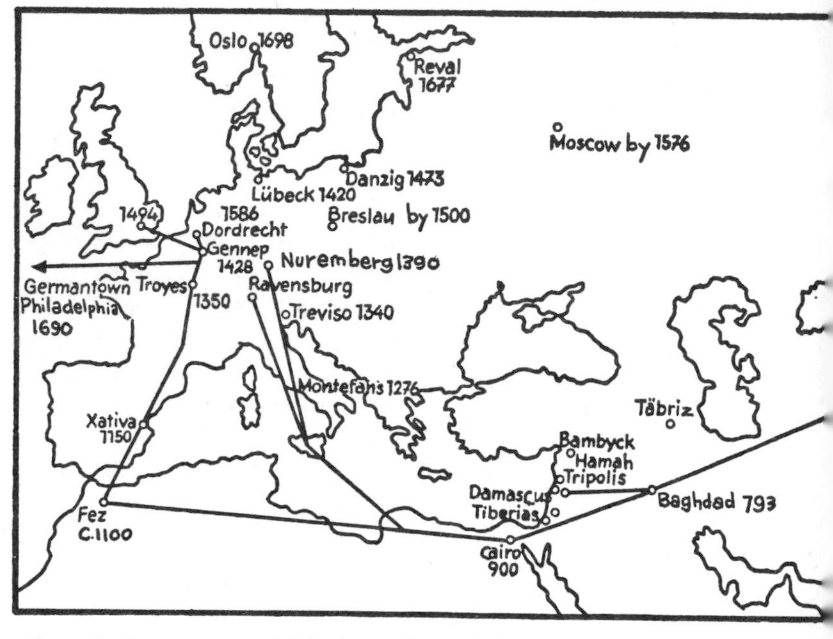

The path of paper westward. The dates refer to the establishment of paper mills. After Armin Renker, Das Buch vom Papier (Insel Verlag).

things, along the Silk Road. But not only did the material travel along this main highway artery of the ancient world, the knowledge of papermaking traveled with it. To be sure the inhabitants of Szechwan tried to maintain their monopolistic position and even succeeded in doing so for a while. But a decisive turn took place in 791 A.D. Samarkand succumbed to the waves of invading Islamic armies. The new rulers recognized the special usefulness of certain Chinese prisoners: they were artisans skilled in papermaking. The followers of Mohammed proved to be generous and farsighted; soon Samarkand was famous for the high quality of its paper. Baghdad and Damascus followed with other paper mills.

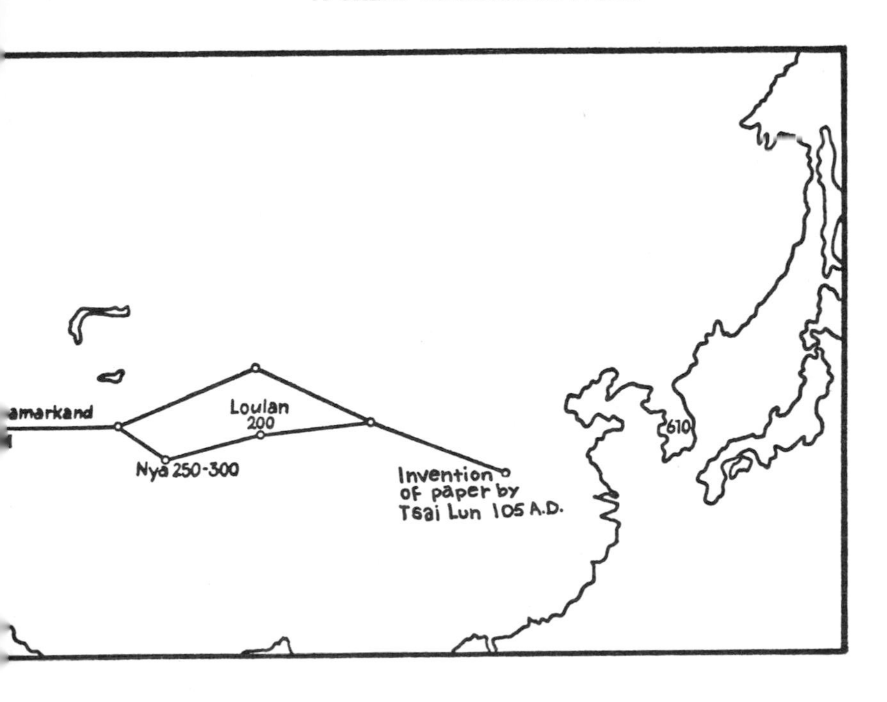

In Egypt the production of paper began around the end of the ninth century and supplanted the papyrus which had been used as writing material up to then. The huge amounts of mummy bindings offered cheap raw materials for the mills in the country on the Nile; grave robbers found significant extra profit therein.

The papermaking craft reached Spain by way of Morocco, where a document from the year 1150 refers to a well-established paper industry. The first paper mills were built in Italy in the thirteenth century, and something was added to the white material that the Chinese did not yet know about: the watermark. The merchant Ulman Stromer of Nuremberg, in

1390, became the first German producer of paper. The diary he left behind is for us today an important document on the method of papermaking in the Middle Ages. But the merchant did not limit himself to such information: he also described the first strike in the history of the paper industry.

Paper is now almost nineteen hundred years old and our daily life without it is unthinkable. Yet once a Chinese scholar excused himself because he used paper for a copy of his work. There is an Egyptian letter of thanks from the year 890 A.D. which closes with the words, "Excuse me for writing on papyrus."

The cheap substitute, paper, was an important invention, but it was the invention of the art of printing that first ushered in great and permanent revolutionary changes which we shall discuss in the next chapter.

The pulp mill of Ulman Stromer before the gates of Nuremberg, erected in 1390. It was the first papermaking plant positively known to have been constructed in Germany. A paper mill in the region of Saint Julien near Troyes was erected even earlier (1348).

9.

The Black Art

THE bells are ringing, and snow crunches under the boots of the inhabitants of Haarlem, who are all pouring into the churches on this Christmas Eve in the year 1437—all except for one person. Apprentice Johannes Faust has feigned illness and is now alone in his master's house. Master Laurens Janszoon Coster is a versatile man: merchant, sexton, and innkeeper. Johannes does not think it is worth his while to look into the cashbox, customarily bare. He is after bigger game. In the workshop are things that will bring him the long-dreamed-of riches. He slides out of bed, slips into the travel clothes that he had laid out the night before, arranges his few belongings in the knapsack and quietly descends the stairs leading to the workshop. The door is not closed and now the metal type lie before him. His employer for some time had been applying himself to the art of printing and had been struck by a brilliant idea. Up until then, tractates and captions for the saints' pictures in wood had been engraved in reverse writing on wood, from which impressions were drawn. Now Master Coster cut up the lines into letters, made a type font of each one, and then set them together in lines and pages in tightly locked forms. He had kept this a secret from others. Only Johannes had been made privy to it. After all, the master had to have an assistant who kneaded the clay, melted the metal for the type, and spread out the sheets of the costly Italian paper in the air to dry.

Yes indeed, Johannes had learned all the tricks of the new craft. Now it was just a matter of stealing the type, escaping over the border, and selling the secret, together with the tools, to another printer.

The knapsack weighs heavily on his shoulders as he leaves the house. The city gates are already closed, but that does not faze him. Johannes knows a side gate. He makes his way to it—an apprentice on his way to the Rhine and farther upstream.

At least that is the way the Dutch historian, Adrian de Jongh, told the story 130 years later in his work *Batavia.* He also purported to know the purchaser of the secret: Johannes Gutenberg, who in the meanwhile has become internationally

A printing shop in the seventeenth century. (Woodcut by Abraham von Werdt.)

famous as the inventor of the "black art." Moreover, the Dutch historian explained why fame had eluded the sexton of Haarlem. After the loss of his tools, Coster simply did not have the money to make new ones and is said to have died in great wretchedness and poverty.

The critical reader will immediately discern the weak point in this argument: why should Johannes Faust have stolen the tools and the type, when he could have made them himself anywhere? The decisive element in this invention was not the toolmaking but the idea of using individual type. Ideas are easy contraband; there was no need for the apprentice to incur the risk of being punished for thievery. Moreover, the oldest fonts of the Master of Mainz show no Dutch influences whatsoever.

But for those who begrudge Johannes Gutenberg's success, of course, these were not (and are still not) considered to be sufficient grounds for consigning the story to oblivion.

Let us turn the wheel of history back several centuries, more exactly to the year 1088 A.D., and let us peer into the study of Shen Kua in distant China. He had just finished another entry in his Encyclopedia and was re-reading the text:

"In the time of the rule of Ching-li, a man of the common people, Pi Sheng, invented movable type. His method of work was as follows: he took clay and cut in it characters as thin as the edge of a copper coin. The same form was given to each character, as though it stood alone. He baked the clay in fire until it hardened. Before that he had smeared an iron plate with a mixture of pine resin, wax, and paper ash. When he was ready to print, he took an iron form and set it on the plate. He placed the characters in it, set close together. When the form was full, he heated it until the paste became soft. Then he took a smooth board and pressed it on the surfaces of the characters until the whole block was as smooth as a mirror.

"If one wanted to print only two or three copies, this method is neither simple nor quick. Yet it is extraordinarily suitable for printing hundreds or thousands of copies. As a rule Pi Sheng worked with two forms. While the apprentices were making the impressions from the one form, he was setting new characters in the other. Thus one page was always ready and the printing was done with great speed.

"There were several pieces of type for each character, and more than twenty for those that are required frequently. Thus he was prepared in the event that a character often appeared on the same page. The type was ordered according to groups and kept in wooden cases. The reason he did not want to use wood for his type, however, was that wood fibers are sometimes fine and sometimes coarse; also wood absorbs moisture and expands in the forms. Wood sticks too firmly to the plate base and great effort is required to pull it out. Thus for him it was more advantageous to use baked clay. When the printing was finished, the forms were heated until the characters could be removed. Up to now the cleansed type cannot be distinguished from the unused type.

"When Pi Sheng died, he left the characters to my cousins." Shen Kua could still vividly remember how, as a youngster, he had watched the master at work. What progress had been made since the beginnings of the art of printing!

The chronicler's gaze swept along the long shelves of his library. It was nothing unusual to possess 50,000 books, but he also owned several rare collector's items. For example, this piece of silk here with the three red lucky signs. A good five hundred years had already gone by since a monk had smeared the characters incised in wood with cinobar, carefully laid the silk over it and rubbed against it with a soft brush. Shen Kua turned the small plate that belonged to it in his hand, reflectively. That was really not yet a printing block, it was more of a seal, intended, moreover, for pressing sacred ideographs in moist clay in order to spread Buddha's message. The walls in the monasteries were bedecked with them, even the entrance to the living quarters protecting the inmates from malevolent spirits.

Who was the first to smear the surface of the seal with color and to impress the characters on pieces of wood, bamboo, silk,

The oldest Chinese printing shop with movable wooden type.

or paper? And where lay the dividing line between the complicated seal and the single printing block? "Yin" in Chinese denotes both seal and print.

Nothing much concerning the beginning of the art of printing had been handed down from the turbulent centuries, Shen Kua mused further. Paper and wood caught fire too easily. The invention of printing was by no means wholly superior to copying by hand. The impermanence of printed products was proved especially in connection with paper. Unfortunately, money was seldom available in sufficient quantities to print huge editions and thus increase the chance of preserving literary work for posterity.

Still, many a printing block was cut in stone and skillful craftsmen could make legible impressions from the original engraving for a very long time thereafter.

An easier method involved cutting a whole number of characters in wood, in reverse writing, from which impressions were made. The astounding number of books that came from the province of Szechwan and spread throughout the Middle Kingdom were not only of a scholarly or religious character. Calendars and horoscopes held out the prospect of a brisk business to entrepreneurs. Now where was that Imperial decree from the Tang period? Shen Kua skimmed through it.

On December 29, 835, the Governor in East Szechwan complained to the Emperor that the calendars for the new year had already been printed and were already being sold in the marketplaces in great number, although the Emperor had not yet approved the new calendar.

The decree forbade the private printing of the calendar. But the desired effect was not achieved: enterprising shopkeepers, under the counter, continued to sell calendars with weather forecasts, prophesies for lucky and unlucky days in the year, edifying sayings, and gems of wisdom.

Shen Kua possessed another precious rarity which had survived the ravages of time: a copy of the "Diamond Sutra." There were many records of the easily remembered teachings of Buddha, but this represented a particularly fine specimen.

The seven printed pages had been carefully pulled from superbly engraved blocks. In keeping with the style of the time, the individual pages were pasted together and formed a scroll over 15 feet in length. The date of its manufacture was noted in the concluding sentence: "Dedicated in honor of his parents and made for universal free distribution by Wang Chieh in the ninth year of Hsien-tung on the fifteenth day of the fourth month (May 11, 868 A.D.)."

The British Museum in London includes a copy of the "Diamond Sutra" among its treasures. It is not the one that Shen Kua called his own in the eleventh century: it comes from the Caves of the Thousand Buddhas. There is an exciting story behind it. In Eastern Turkestan, in the proximity of the city of Tun-huang, of no importance today, the Silk Road once crossed the ancient trade route that led from Siberia to India by way of Lhasa. A countless number of caves have been excavated in a wall of rock not far from the hustle and bustle of the Silk Road. Monks had lived in these habitations since the fourth century. In the year 1900 the stream of commodities flowing along the Silk Road had for a long time shrunk to an insignificant trickle and Tun-huang was reduced to a desert town, forgotten by the rest of the world. Only some wretched inhabitants still lived in the caves. There was little about the setting that recalled the brilliant past of what had once been an important religious center. Only the activity of a Taoist monk disturbed the peace that had descended upon the locality in its decline. He busied himself by restoring the frescoes of a cell to a new brilliance. He had gone around begging all his life, and had saved his alms for this purpose. Now he had progressed so far in his labors that the faded ancient frescoes were ablaze with new colors, the debris had been removed from the cell and he had also repaired some broken sections on the walls. During this latter operation he discovered that the rear wall of the cell was not all rock. Broken plaster lay on the ground and brought brick to light. Under the monk's probing pick, the bricks fell to the ground and disclosed a chamber piled up to the ceiling with scrolls. The secret chamber contained 1130 bundles carefully sewed up in cloth. They had been marvelously preserved by the desert climate.

We can only assume what had happened. Danger must have threatened the cave city so that in the year 1035 A.D., the monks had brought all writings to safety behind a secret walled-up chamber. Knowledge of the existence of this rich storehouse remained lost for almost 900 years. And it was indeed a treasure trove: not only did the "Diamond Sutra" and other Chinese records come to light, but Tibetan, Persian, and Turkish writings were also found among them. The investigators even discovered a book with selections from the Old Testament among the more than 15,000 texts. All stem from the time between the fifth and the tenth century.

But things had not yet progressed so far. The involuntary discoverer, Wang Tao-shih, had only a dim awareness of what had come to light behind the wall. The times were bad, robbers infested the countryside, and every honest man managed to earn his daily bread only with the greatest difficulty. If the Governor were to get wind of the find, he would sell it to foreigners for his personal gain. And he would certainly throw Wang Tao-shih into prison as a dangerous accessory. No, it was advisable to wall up the find again.

Several years later Sir Aurel Stein and his expedition were riding along the abandoned Silk Road. He also visited the Caves of the Thousand Buddhas. There he heard rumors about Wang Tao-shih's treasure. He asked to see him. The monk modestly denied knowing anything of any hidden scrolls. But Stein pursued his inquiry adamantly; the scrolls would crown the success of the expedition. And even if the rumor merely approximated the truth, this find was bound to be quite different from the few paper fragments that he had found in the ruins of the Great Wall. He drew on all powers of persuasion at his command, emptied many cups of tea in conversation and slowly, very slowly, won the confidence of the saintly caretaker. Sir Aurel Stein's account of his investigations concerning the wandering monk Hsüan Tsang settled the matter. Wang Tao-shih also revered the restless God-seeker;

whoever was so well-informed about the venerable pilgrim, he felt, could not have anything up his sleeve.

On the next evening he brought an armful of bundles sewn in cloth. Tensely he watched as the foreigner grabbed one and carefully cut it open. Sir Aurel Stein took out one of the scrolls emerging in the light of day, opened it and after a brief look at it, he handed it over to the monk for his inspection. It contained a biography of Hsüan Tsang whom they both revered!

This was a good omen, and the number of bundles removed from the hiding place grew from day to day. The investigator packed his scientific booty, consisting altogether of more than three thousand scrolls and some specimens of silk, likewise discovered, in 27 cases. The storehouse was still far from emptied when the expedition departed with a heavy heart. The monk remained behind. His misgivings over the propriety of his action had been allayed by the presentation to him of Sir Aurel Stein's entire capital in silver.

Now he could renovate the cells even more beautifully. But that was not the only project that he undertook with the unexpected wealth. He planted many hundreds of trees and had wells dug in the vicinity. He then built a hostel on the site of the oasis which he had created in which he accommodated weary pilgrims and travelers. Travelers still treasure the hospitality they enjoyed there in the 1930's.

When news of the precious finds spread in Europe, archeologists wasted no time. A year later Professor Pelliot also purchased three thousand manuscripts for France. Thereupon the local authorities pricked up their ears. The remainder of the scrolls were confiscated without payment. And Wang Tao-shih often had occasion to regret that he had not handed over the entire find to the friendly Englishman who was so knowledgeable about the Tao. The scrolls seized by the Governor were handled with scant respect and a great part of them were lost during transport.

A few years later Sir Aurel Stein also discovered in a Buddhist monastery a paper fragment assertedly containing the oldest printed inscription. This paper, dated 594 A.D., which achieved world-wide fame, was translated in the 1930's by Professor Maspero as follows: "This house has a snappish dog. Passersby beware!" If not true, it is cleverly fabricated, as the Italian saying goes. The translation is incorrect, unfortunately, and the fragment is not printed.

Printing with movable type did not find proper acceptance in China. In order to understand this, it is worthwhile briefly to consider the Chinese written language. Originally Chinese was an ideographic language, yet today only 5 per cent of all the ideographs are simple symbols; the vast majority have been combined to create complicated concepts. Despite many reforms–recently the Chinese Peoples' Republic again simplified 1000 symbols and introduced the style of writing from left to right–it is the only ideographic language in the world that has substantially remained unchanged for more than three thousand years. An educated Chinese can read and understand most of the two thousand characters of the Shang period (c. 1520-1030 B.C.). What that signifies becomes clear when we think of the difficulties a German, say, would have in reading the "Nibelungenlied" in the original version–which stems from the thirteenth century and therefore is only seven hundred years old.

The Chinese script contains around 50,000 characters, but only 412 different sounds are available for their articulation. Each sound therefore has many meanings and the sense of the spoken word can be derived only from the context of the completed sentence. Since countless dialects are spoken in a country as vast as China, there are naturally problems of understanding one another among the inhabitants of the different regions. But this is only true for oral communication. The meaning of the characters is uniform in every corner of the Middle Kingdom, even the most remote.

If we want to learn the pronunciation of a current word symbol we must turn to sound tables that show this word along with similarly pronounced words. The exact phonetic value can be taken from the position of the word in these tables. This is somewhat similar to the way we interpolate intermediate values from the neighboring values of a logarithm table.

The written language, comprehensible to all Chinese, played–and still plays–an important role with respect to the formation and preservation of Chinese national feeling. Whether the Chinese land was split up into individual partial kingdoms or lay under foreign domination, the script was a common clamp that held the nation together at all times. Now it is also clear why printing with movable type did not find a special acceptance in the Middle Kingdom. The supply of type fonts required would have been truly enormous. Thus it was much simpler to make one's own block: wood panel or block printing were more economical.

Now and then the State undertook to attack the problem–a project of this kind would have been beyond the means of private persons–and manufactured type in great number out of wood. Thus we hear of 360,000 type fonts from the Middle Ages, some of which can still be seen in the museums. Nevertheless, block printing prevailed over any other form of printing. The complete book page was cut into a piece of wood.

The process was quite simple: the calligrapher wrote the text with bold ink on extremely thin paper; the page, still freshly covered with writing, was pressed over the smoothed piece of wood and left its impression behind in reverse writing. Expert engravers cut out the symbols. If an error had slipped into the text, or a part fell out, or if the block wore out, a new one was quickly reproduced.

In practice it is very hard to distinguish whether a discovered book has been produced by means of block printing or letter press. In China wood primarily served as the initial material, and the end product no longer betrayed the printing technique–except for a few exceptions. There are texts in which a character lies on the side, a printing error. And that, of course, could happen only if the printer inserted an individual character (type) in reverse in the chases.

Printing by means of movable type, therefore, was already

known in the middle of the eleventh century in China. Nevertheless, the archeologist Carl Hentze believes to have found this process on bronze casts from 300 B.C. From many small clues he concludes that the inscriptions cast in bronze were put together from individual characters. Yet even if his proof is valid, we should not call it printing by means of movable type. One essential feature is lacking: the process was not used to duplicate the written word.

Even if many works were printed by means of movable type, this process never achieved great importance in China. By the nineteenth century even the knowledge of the process was almost lost.

"Dokyo, why does anxiety never leave me? What can I do so that Buddha may send me peace and help me forget the terrible times past so that my heart is not torn apart a thousandfold by the fear of death?

"The priests of the Enlightened One find my court hospitable; you yourself, his High Priest, are a king in the eyes of every man. We have cast his form in bronze, the sound of the new mighty bell reaches him every day. What, Dokyo, can I still accomplish in order to assure myself a long life?" Shotoku, queen at the Japanese court in Nara, gazed upon the High Priest questioningly.

"Listen to a parable out of the Sutra, O Queen: a sick Brahmin in his need turned to a seer. 'You will die in seven days,' he was told. So he betook himself to the Buddha and pleaded with him to save him and to receive him as a disciple. The Enlightened One said to him: 'A pagoda has collapsed in a certain city. Go thither and rebuild it. Then write a charm and leave it behind there. The reading of this charm will lengthen your life and assure your place in Paradise.' The disciple asked the Master wherein lay the power of the charm. And Buddha answered, 'Whoever wants to derive benefit from the charm must write it down 77 times and keep the copies in a pagoda. The pagoda must then be honored with sacrifice. But one can also make seventy-seven pagodas out of clay and leave a charm in each of them. The life of the benefactor will be saved and his sins will be forgiven.' Reflect upon the parable, O Queen," concluded the head of the Buddhist priesthood who, at that time, was the most powerful figure in Japan.

And the Queen, as always, followed his advice. In her early youth she had been saved from a raging smallpox epidemic on the very edge of death. Now she had but one aim: to live as long as possible. In order to dispose the fates kindly to her, she had surrounded herself with priests and completed the erection of a huge statue of Buddha. It was some 50 feet high and weighed over five hundred and fifty tons and can be seen today in Nara. Here another way had been offered to her for obtaining the Gautama's grace.

She wanted to proceed with a built-in guarantee that her operation would be conducted on a sufficiently large scale, so she ordered the priest to publish a million charms from the Sutra. Thus was mass printing born.

The succinct, easily remembered precepts in Sanskrit were transliterated and the Sanskrit sounds were reproduced as nearly as possible by Chinese characters and engraved on copper plates. Although anybody could read the precepts, they remained incomprehensible—but that was of no matter. The same held true in the West for many hundreds of years. Church communities pronounced Latin prayers by rote and understood not a word of them.

A total of 150 characters were cut in one printing block. The paper charm was about eighteen inches long and two inches wide. The imperial workshop built a miniature wooden pagoda, four and a half inches high and three and a half inches wide at the base, for each sheet, which served as a mini-shrine. They were distributed to all the temples in the country and preserved there. Some have withstood the vicissitudes of time and today are preserved in museums as precious possessions.

Is not the first big printing order in the world very impressive? To be sure, printing had taken place before, especially in China, but never in an enormous edition of this kind. The problems of mass production cropped up: each printing block had to be prepared several times because of the rapid wear and tear. Paper production enjoyed a thriving business, and thought had to be given to details such as whether sufficient supplies of ink were on hand.

Four different charms were chosen, transliterated from Sanskrit into Chinese characters in Japan. The task was accomplished after six years, in 769 A.D. But the Queen's expectations, alas, were not fulfilled, since she died in the very same year.

The successor to the throne in Nara failed to break the influence of the Buddhist priests over the affairs of State. But he finally shook off their domination with a brilliant idea and became one of the most important emperors of Japan: he commended the shrine of Nara to the care of the priests and transferred the seat of his court to Kyoto.

Another half century went by before the next most important printing order in history was placed. If the Japanese Empress had been concerned about the length of her life, the Chinese Chancellor Feng Tao (882-954 A.D.) was beset by quite different problems. In the midst of the troubled times that marked the third partition of China, he himself served as the highest official of ten different emperors from four Chinese and one Mongolian dynasty, and he wanted to fix a period of respite, a symbolic act that would also be recognized by later generations. So he commissioned the scholars of the Imperial Academy to collect the writings of Confucius, to sift them critically, and to approve a standard text. Thenceforth their reaction was to be viewed as the only authentic edition of Confucius.

Feng Tao, whose services so many crowned, short-lived heads found non-expendable, always planned with a view to the future. He would have preferred the collected and corrected text to be engraved on stone plates. This process had been known for centuries and led not only to permanent inscriptions, but it was also possible to make impressions from them.

The State coffers, however, were bare, and the Chancellor had to cope with countless other projects involving huge outlays of funds. Consequently, he ordered that the final text be cut into blocks of wood from which books could then be printed. Perhaps a large edition would prevent the work of the Academy from being lost in times of trouble.

And this is precisely what happened. While on all sides the country was being sucked into the vortex of struggles for succession to the throne, cities and villages were being set to the torch, and one ruler followed another in rapid succession, the Academy finished its work in utter peace and quiet after twenty-one years. The importance of this printing can be compared only to the 42-line Gutenberg Bible. Feng Tao is therefore honored in China as the real inventor of printing.

The invention of printing had a revolutionizing impact on the development of the West. Now for the first time the written word could be brought to the masses. The education monopoly of a thin stratum collapsed and thus in the long run led to mass education, a prerequisite for democracy.

Not so in China. Nothing changed in the structure of society. Democracy already prevailed within the educated stratum in the sense that the choice of the ruling officials was made from among the educated officials. Printing merely broadened the circle of candidates slightly.

Under the Sung dynasty (960-1234) the art of block printing achieved a peak of perfection never to be matched thereafter. The best calligraphers of the Middle Kingdom competed with each other in the preparation of the copy for the wood carvers who incised the complicated characters on smoothed pear-tree plates. Only the finest qualities of paper were deemed good enough and the printers also exercised great care. Strictly speaking, we should not call it printing since it was, rather, a process of obtaining impressions. The sheet of paper was placed on a block covered with printer's ink and gently gone over with a soft brush. Gutenberg was the first to press the paper on the printing block.

Today we not only admire the quality of the printing of that time, but the quantity also commands our respect. Thus, for example, a commentary on the works of Confucius was published in one hundred and eighty volumes. The Triptaka, the canon of Buddha's writings, appeared between 972 and 983 A.D. in 5048 volumes, totaling 130,000 pages. Each individual character of each page was cut by hand. And we can imagine how many printing blocks had to be prepared over and over again since wood quickly wears out.

The first private publishers arose in this period. They published for profit not only canonical texts–as did most of the State printing establishments–but also short stories, poetry, historical literature, scientific manuals, and, as we already know, calendars. The accounting concerning a book put out by one of these private publishing houses in the year 1176 has been preserved:

"Paid to the printer for paper, paste, and labor:
Per copy .... 1500 cash*
Sale price per copy .... 8000 cash
Rent of printing blocks .... 1200 cash"

Eight hundred years ago one could already earn a good living from books. Unfortunately, we do not know how large an edition it was so that present-day publishers could make an interesting comparison.

Printing shops were established everywhere in the Chinese sphere of influence in the early Middle Ages. Yet a work like the Great Encyclopedia of 1403 was too extensive even for this efficient industry. If the volumes were set side by side they would extend some 425 feet. It was not until 1567 that scribes prepared two copies of it. Only 368 volumes, scattered in the museums of the whole world, have come down to us. Two complete sets were lost in Nanking in 1644 during the struggle against the Manchus, and the rest burned in Peking during the Boxer Rebellion in 1900. We would not have needed to append question marks to so many passages in this book if the Great Encyclopedia had been preserved.

**Cash* is a Chinese copper-alloy coin. –Trans.

Meanwhile, in the West, monks sat in their monasteries and prepared hand-written copies. It was hard and painstaking work. The knowledge of printing did not reach Europe. Why? The Islamic Empire bolted and barred off Europe. We do not know exactly why, but it is a fact that in the empire of Mohammed the art of printing was forbidden. Perhaps it was due to the stubborn adherence to once meaningful traditions. The Koran could be copied only by hand. It was not until 1825 that the first book printed by Moslems appeared in Cairo.

Thus it was reserved to Gutenberg to develop the art of printing independently of the knowledge acquired by the Chinese. Whereby it behooves us to note that the son of the Mainz patrician not only discovered printing by means of movable type anew, but also led printing technique to a peak of perfection. The movable type from Korea is not uniform and is more primitive than those used at the same time in the first Bible. Gutenberg succeeded in casting the type so perfectly that the pieces of type formed the printing block as such by being placed directly alongside each other. The printing press is also his invention.

One Chinese printed product, however, was very well-known in medieval Europe: as early as the tenth century playing cards had made their way to the West along the Silk Road. What does the oldest preserved printing block (1377) in Germany depict? A playing card!

We shall not, however, be turning our attention to this in the following chapter, but rather to another printed product that many people consider to be indeed the most important of all: paper money.

Title playing cards from China, eighteenth century. Four colors represented the classes civilian, military, scientific, and academician. Nine values corresponded to various titles such as Minister, Governor, or Treasurer.

10.

Paper Money, Inflation, and Currency Reform

Mephistopheles:

> Such paper, stead of gold and jewelry,
> So handy is–one knows one's property:
> One has no need of bargains or exchanges,
> But drinks of love or wine, as fancy ranges.
> If one needs coin, the brokers ready stand,
> And if it fail, one digs awhile the land.

*(Faustus II)**

DOES one really know one's own property? Few readers would agree with Mephistopheles, whereas they would have an instant and sympathetic understanding for the following lamentation:

"After they had esteemed and used paper money for years, people lost confidence in it, indeed, they were afraid of it. For the government made its payments exclusively in paper. The

*Trans. by Bayard Taylor, New York, 1967.

output of the salt mines was changed into paper. The salaries of the officials consisted of paper. The Army received its pay in paper. The provinces and districts, up to their ears in financial difficulties, paid their debts in paper. Copper money–it was seen but seldom–was guarded like a treasure. No wonder then that the cost of living rose and the value of paper money dropped. People lost all desire to work. The soldiers did not know how they would fill their bellies on the morrow. The inferior officials of the empire did not even have enough to procure the common necessities. All this was the consequence of the depreciation of paper money."

Does all this have a familiar ring? We can read that kind of a complaint every day in the newspapers. Perhaps it may be a comfort to know that the citation comes from the pen of the Chinese historian Ma Tuan-lin. It was written at the beginning of the thirteenth century.

Then did paper money already exist seven hundred years ago? Strictly speaking, the birth of this invention, which had such far-reaching consequences, occurred in the year 994 A.D.

Some may object that this is not so. Paper money has existed ever since the invention of paper. Yes and no. This kind of paper money was of a purely symbolic kind and was never in circulation as a means of payment. It was money that was placed in the graves along with the dead so that they would not be without means in the world beyond. Furthermore, this spirit money exhibited no resemblance whatsoever to actual coins. Compared to gold and silver it had the advantage that grave robbers could not derive any benefit from it and this took all the pleasure out of their activities. From the fourth century on, it more and more became the custom to burn this "money" and forward it thus to the deceased. People placed such confidence in the spirit money that they gave up all other grave gifts. After all, everything could be bought in the world beyond if one had merely burned enough money.

But its use for grave gifts does not really mark the beginnings of the history of paper money.

Actually it came into being more through accident than

intention. No Mephistopheles whispered the idea into the Emperor's ear: the mint of the province of Szechuan could not procure for coinage enough metal, whether silver, copper, or iron. The confusion and disorder of the times had provoked an uprising against the Emperor. It had just been put down; trade, traffic, and handicrafts lay prostrate. What to do? The populace needed an exchange medium. The supply of coins in circulation shrank, and the demand rose constantly. Domestic trade in the unified Empire of the Sung dynasty was flourishing–apart from occasional, local setbacks–and that required more and more currency. Chang-ho, who was the official in charge of coinage, was at his wit's end: "When the merchants deliver goods to the Emperor, they shall receive a certificate denoting the value in silver. This certificate can be presented to any provincial treasury and exchanged for currency. We all know it as chia-tzu, or credit money. Let us print chia-tzu until times improve and we again have a sufficient supply of metal coinage."

His suggestion was acted upon. Scribes inscribed a value on the certificates, and the competent official authenticated it with the seal of the province. The district treasuries then put them in circulation in the place of iron money. This paper money was expressly considered merely as a temporary expedient and was supposed to be converted into metal currency after three years.

It is reported of Emperor Wu (120 B.C.) that he had rectangular pieces of white deerskin artistically painted and sold them at exorbitant prices to the nobility of the country, putting a little pressure on them to buy. The nobles could do naught else. Then in the following year the Emperor subtly suggested to them that they should present the leather pieces to him as a tribute so that he could begin this game all over again.

Thanks to this infamous method, the Emperor not only filled the State coffers, but at the same time reduced the influence of the nobility and thus came closer to his goal of centralizing power in his hands. He had no need to fear that skilled craftsmen in the service of the landed aristocracy would imitate the painted leather pieces: the deer whose skins could be processed into white leather lived only in the park of his palace. The animals were considered sacred and only the Emperor enjoyed the privilege of hunting them.

We can hardly call this leather money: no one accepted it as counter-value in payment for goods. Indeed many readers may exclaim: "But that reminds me of the confounded modern currency system." Not so fast! The analogies to the twentieth century become even more striking. For the great potential that lay in paper money did not remain hidden for long.

First came the forgers. The printing establishments required large supplies of paper, and therefore paper became a precious commodity to own. Special paper with watermarks and worked-in metal threads were refinements reserved for later centuries. Any skilled wood carver could imitate the seal of the province. It was not difficult to manufacture money and soon, therefore, huge amounts of bogus money were in circulation.

Acute officials also saw the possibilities: before long more paper money left the treasuries of the provinces than one could hope to change back into metal after three years. An inventive mind found the solution: after three years, the bearers of the old notes could simply exchange them for new ones. Finance Ministers ought to erect a monument to this Chinese whose name, unfortunately, is unknown to us.

That could not go on for long, and consequently the value of the paper money fell–nobody accepted it any more for its designated value in payment. After some fruitless attempts at intervention by far-seeing officials, the central government took the matter in hand and issued stringent decrees on the issuance of paper money and its conversion to iron, copper, or silver. It is interesting to note the date of this decree: April 1, 1024. Paper money was now considered a valid currency; it had established itself as a means of payment everywhere in China.

We can confidently assume that the notes were now printed, since the demand could hardly have been met with handwritten notes. The oldest preserved references to a copper printing block for the printing of notes, however, stem from the year 1163.

Only the next two generations of Chinese enjoyed a stable paper currency, then a rapid decline set in. The impulse for the inflation that began toward the end of the eleventh century came from outside the country.

Countless bands of horsemen from the steppes in the north and northeast of the Empire surged against China. The Sung had neglected the Great Wall, the bulwark was in a sorry state. Nor were things much better in the Army, whose weapons were supposed to hold the enemy in check from the heights of bastions and battlements. Educated Chinese, especially the officials administering the State, held the profession of soldier in contempt. The man of rank and esteem wielded a writing brush, not a sword. This spirit also infected the troops and reduced their combat capabilities.

On top of this, individual barbarian peoples had been allowed to settle within the perimeter of the Great Wall. The Empire paid them subsidies, and thus hoped to establish a buffer against the outside world. This was an enormous miscalculation; instead, the invited tribes pitched their tents within the Wall and held their ground against all attempts to dislodge them.

One region after another fell victim to the Mongolian drive for conquest. Payments of tribute at least bought a temporary armistice and postponed the final defeat of the Sung. The reparations, still called tribute at that time, obviously could not be paid in paper money. Just as the Allies, after the two World Wars, carried off whole factories, ships, patents, railways, and other hardware, so did the Khans demand the riches of their time: gold, silver, silk, pearls, and precious stones. And the defeated governments, in both cases had recourse to the same means in order to pay off the oppressive indemnities. They let the note-printing presses run at high speed. Hence it is no surprise that in 1107 the first currency devaluation, a currency reform in the ratio of 4 to 1, was announced. A few years later the population had to exchange its paper money again, this time 10 to 1—against new paper money, of course.

Although the notes were not worth very much, the government made great efforts to make them appear valuable. The money looked more magnificent, the paper used was more expensive. One of the high points of this development is reflected in an account dating from 1209: perfumed notes made of silk paper promised to pay the counter-value in gold as soon as times improved. As customary, the owners of these notes also waited in vain: in 1260 Kublai Khan completed the conquest of China.

The victor had already taken over the money system which was so handy. Paper money showed its positive properties in the solidly administered giant empire of the Great Khan. It facilitated the flow of commodities within the vast domain and was also accepted without the least hesitation by the merchants of neighboring States. Confidence in the greatest empire of all times secured the paper money and maintained its value.

In Marco Polo's time merchants called paper "Charta Bambycina," paper made out of bamboo. The error began with the bowdlerization "Charta Bombycina," paper made out of silk. It was assumed that the Arabs–as the European purveyors of paper–replaced the expensive raw material silk by cotton, but charged their customers for silk. Actually from the outset paper was also made out of rags, a process that up to recently was credited to the papermakers of the late European Middle Ages.

At the time of the Ming dynasty (1368-1644 A.D.) money was printed on gray paper. Why? Emperor Wu (1368-1398) had asked his advisors how paper could be made so as to foil forgers. As a solution the learned Mandarins recommended that the hearts of the literati be pounded into a pulp and added to the raw materials of which the paper was fashioned.

The Emperor, however, did not like the idea of losing his writers and sought advice from his consort. She came up with a compromise: "The heart of the writers lives in their works. Let books written by them be pounded into pulp and make paper money out of it." The black ink of the characters colored the new note paper a dark gray.

Presumably paper money is the first printed product and

probably the only one–aside from playing cards–of which Europeans had heard about before Gutenberg's time.

It took a long time before paper money was printed in Europe. As in China it remained valid and readily accepted currency as long as a politically and economically strong government stood behind it. The value of a paper currency dropped with the power of the State that had issued it.

Some readers will "thank" the Chinese for the ingeniousness to invent paper money but point to the good old days of gold currency. It may be a comfort to them to know that the ancient Chinese themselves experienced all the effects of paper currency, including a ruinous inflation and currency reform.

11.
Gunpowder and Cannons

THE ring had closed, the capital of the Sung dynasty, Kaifeng, looked forward with great trepidation to the siege that was just beginning (1126 A.D.). The watch fires of the Juchen Tartars flickered at a respectful distance from the city walls. They had come from very far away, from the mountains and steppes of Mongolia. The tribes, otherwise so splintered, had united into a powerful association. In the fertile highlands, plains, and river valleys of China they had glimpsed the Promised Land and the possibility of nourishing the constantly growing number of their kith and kin. For they had not come as thieving aggressors, in search only of booty with which to return to their camp. Their aim now was to conquer a new homeland for themselves.

In the beginning their push proceeded slowly. They first had to learn how to break down the walls of solidly fortified cities. These horsemen and shepherds, however, had learned quickly and now they stood before the last bastion of the Sung. If Kaifeng fell, the new lords, the Chin—they called themselves "The Golden Dynasty"—would take over as the successor of the Sung in northern China.

A well-equipped arsenal was at the disposal of the defenders behind the walls. It is a definite fact that the scholars and officials of the Sung dynasty detested the profession of arms and dedicated their attention primarily to pure science such as astronomy, mathematics, and philosophy. Nevertheless, they had not wholly neglected weapon technology and the commanding General Yao Yu-chung had many means for the defense of Kaifeng at his disposal.

The General had decided to use *huo pao,* the fire drug, as often as possible; it was said to have a shattering effect on the morale of the Juchens. Unfortunately, this versatile weapon was no longer unknown to the enemy, since other cities had already used it earlier. But this time it was not to be just one among a number of means of defense. It was to become the main weapon of the defenders of Kaifeng.

Yao Yu-chung had made all the necessary preparations, the raw materials lay ready before him. He could begin with the dangerous procedure of mixing the basic substance. He leafed once more through the pages of Wu Ching Tsung Yao's military manual. It had been assembled 86 years before (1040 A.D.) and contained the accumulated wisdom of many generations: "Crush and mix one pound, 14 ounces of sulphur, 2½ pounds of saltpeter, 5 ounces of charcoal, 2½ ounces of pitch, and 2½ ounces of varnish. Then stir a paste out of 2 ounces of dried leaves, 5 ounces of oil, and 2½ ounces of wax. Now combine all these substances and mix them carefully. Keep the finished mixture in a wrapper consisting of 5 layers of paper, tie it with hemp rope, and pour pitch and wax over the package."

The General checked the powder formula against the depot lists. He nodded in satisfaction; everything was available in the required amounts. Paper for wrapping the finished fire drug was hardly necessary. The powder would not age, nor was the protection against moisture prescribed in the manual necessary.

Now he turned his attention to a bundle, wrapped in red silk, that a servant had just set cautiously on the floor. He weighed it: about five pounds; the ballisticians would be able to catapult it over a long distance. "The alchemists will have nothing to do with us soldiers, but when their own hides are at stake, they also concern themselves with useful matters," mused Yao Yu-chung. This missile represented the last improvement of the stink bomb that had been known since time immemorial. This time a surprise lay in store for the foe: the catapulted missile would not only spread an intolerable stench, poisonous gases were also supposed to pour out of it. The smoke screens were not lethal, as the alchemists had informed him yesterday; casualties would merely bleed profusely from the mouth and nose. If the winds were favorable, the General planned to venture sorties behind the protective screen offered by the poison gas bombs and inflict severe losses on the Juchens.

He unfolded the silk: paper, pitch, and stands of hemp hid the actual content of the stink bomb. A silk sheet listed the ingredients in the composition: 15 ounces of sulphur, 1 pound, 14 ounces of saltpeter, 5 ounces of dried aconite tubers, 5 ounces of pulverized soya beans, 5 ounces of langtu, a poisonous plant, 5 ounces of oil, 5 ounces of charcoal, 2½ ounces of pitch, 2 ounces of arsenic oxide, 1 ounce of beeswax, and one ounce of bamboo fibers.

The General shrugged his shoulders in disgust. What ugly expedients people resorted to today! Imagine, poisoning the enemy from afar instead of an eye-to-eye confrontation with lance and sword. Yet even the old war horse had to admit it to himself: the age of chivalry was irretrievably over. Now it was a question of employing cunning and trickery if the empire was not to succumb to the predatory and aggressive fury of the Tartars. Leather doublet and helmet lay ready at hand. It was time to inspect the walls.

The situation looked bad on the northeast wall. The heaviest siege-catapult concentrated its projectiles here. The thick curtains of linen and leather hanging from the endangered section offered but little protection. This method stood the test against battering rams but hardly against the projectiles of the enemy ballisticians. Individual stones had already been dislodged from the wall, and it would not be long before a breach would be made in the rampart. Where had the Juchens gotten that many modern siege machines anyway? Rumors circulated that they had in their service clever engineers from countries lying far in the West, and there was talk of fabulous remunerations being received by them. Yao Yu-chung swept these thoughts brusquely aside. What could these barbarians know! Besides he had not yet seen a device in the enemy camp that was unfamiliar to him. To be sure the Tartars were known for their adaptability; presumably they had reproduced captured catapults and forced Chinese prisoners to operate them.

"Are the *meng huo yu,* the fire-oil shooters, ready?"

"They are over there, in the niche, the fire oil is poured in and the detonators are covered with *huo pao.*"

The General turned to the flame-spurting machines. From the rectangular supply tank a pump transported a special oil into the horizontally laid-out belching tube. Fire-drug paste closed the opening. Should the enemy try an assault on the breach, the gunner would ignite the powder. This would then set aflame the oil waiting in the tube, and the action of a piston rod caused the magic fire to be shot forth amid the enemy ranks. The General further assured himself that it was a double piston rod. The velocity attained by the piston-rod action could be decisive in this endangered section of the wall. It was a tested weapon, which had been introduced in the Military Academy over 200 years ago. Formerly the oil had to be ignited by means of a fuse and other easily flammable material. This result was now achieved more simply and more rapidly with powder. The fuel oil gushing forth from the ground in Kansu and Szechuan did not ignite easily, but once it started burning it could no longer be extinguished. Moreover, voluminous bellows stood ready which would blow the smoke and flames into the ranks of attacking troops. Woe to the soldier who was hit with the flaming oil.

With the *meng huo yu,* the fire drug ignited just the oil. But there was still another, more rarely used type of flame thrower. A hollowed out segment of a bamboo, closed at one end and so thick in diameter that the gunner could put his whole fist

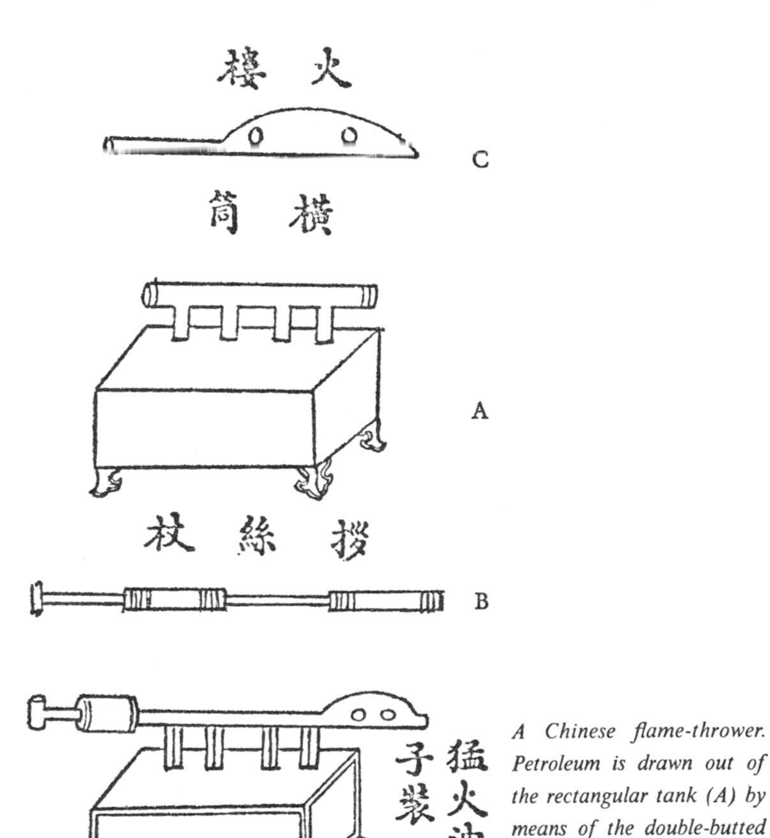

A Chinese flame-thrower. Petroleum is drawn out of the rectangular tank (A) by means of the double-butted pump (B) into the flame-projector tube, and ignited with the "fire drug" in the spout (C).

inside, was primed with powder. Fire from the outside ran inside along a fuse and caused the powder to burn with an audible sputtering. The jet of flame shooting out of the tube burned for about five minutes and covered a distance of several yards. If enough of these flame throwers were available, a breach in a wall could be successfully defended for a long time, until a hastily erected stockade wall fended off the danger.

At the endangered wall, all preparations had been made to meet the expected assault. The General climbed the broad steps to the rampart of the northeast tower in order to inspire the distance-record-setting ballisticians. The tower shook under the impact of each projectile that shot out of the cumbersome catapult. The roaring noise produced by the lever arm was immediately replaced by the screeching in the wind. Here stood one of the many flame throwers. For a century now the generals of the Sung dynasty had learned that such weaponry was not limited to hurling heavy rock and pointed tree trunks into the enemy camp. Yao Yu-chung, beaming with satisfaction, eyed the movements of the soldier who was laying a clay projectile, smeared with a thick layer of *huo pao* on the arm of the catapult, changing it into a sputtering and smoking incendiary shell with a burning piece of oakum, after which he pulled the release. The flaming trajectory of the projectile ended up on the main enemy camp and set a tent afire. The commander-in-chief was lavish in his praise and promised the lieutenant in command an early promotion. Toward noon containers filled with flammable oil that could be used as incendiary bombs were also delivered to the spot. The defenders on the walls were already equipped with these hand grenades and were only waiting for the Juchens to come within the range of fire. Crossbowmen and archers supplemented the catapults. They dipped the points of their arrows in *huo pao* and ignited the fire drug immediately before shooting them. The repeating crossbowmen especially proved themselves to be very effective against mass assaults; one of these skillful shots could replace three to four ordinary defenders.

Now the General directed his attention to a paper kite, which, rising up from the Tartar lines, was being driven by the wind in the direction of the city wall. It hovered in the air directly over the already badly damaged northeastern section. What could that mean? What did the enemy have up its sleeve? The kite could not be carrying an observer because it was too small for that. Were the city's defenders perhaps supposed to be terrified by the painted caricature of a dragon? Hardly. They had all flown kites as children. Something was happening up there which for the moment could not be determined.

Messengers from the other defense sections arrived quickly. There the besiegers had likewise set up flame throwers and already two of the wooden watchtowers had been burned to ashes. Their missiles were tipped with hooks and stuck to every tree trunk. Even portable siege towers had been observed from afar; they were much higher than the ramparts. The attack could be expected early in the morning. Moreover, the Mongols had hurled bundles of leaflets into the city which called for the assassination of the commander-in-chief. The assassin was promised a reward of 10,000 *cash* and a safe-conduct pass.

Yao Yu-chung felt honored by the amount of the reward but he attached no importance whatsoever to the threat. "Drop a similar leaflet on the enemy camp. Say that I will pay double to him who delivers the head of the Tartar chieftain to me."

Meanwhile the engineer Wang Ling had learned the meaning of the kite: the Tartars were digging a tunnel in the direction of the walls and they were calculating the distance between the tunnel entrance and the wall foundation from the position of the kite. Judging by the sounds, the head of the tunnel would reach the wall toward midnight. Enormous containers, the mouths of which were spanned tautly with animal skin, were buried in the foot of the wall. Thanks to the help of this "microphone" the defenders were kept informed on the progress of the tunneling operation.

Yao Yu-chung remained calm and relaxed. Although the situation looked bad, he had not exhausted all his means of defense by a long shot. He would prepare a nice little surprise for the tunnel builders.

Under the protection of darkness porters slipped through the wicket gates. They heaped load after load of powder on the place designated by Wang Ling. This presumably was the point at which the diggers would reach the foot of the wall. Others carried clay, pitch, sand, and stone to the same spot. The engineer himself led the workers. First of all several hundred pounds of precious *huo pao* were piled in a heap, over which a dome of sturdy wood was erected. The fissures were stuffed with pitch and clay, after which sand, clay, and stones were placed in layers over the mound.

Wang Ling was the last to withdraw to the wicket gate shortly before midnight; a track of powder connected him with the mound, barely visible in the darkness. Everything ready? The General nodded, Wang Ling ignited the powder track, the flames raced along it sputtering and hissing.

Just like fireworks, thought the assembled alchemists of the Imperial Academy. And what fireworks! The earth trembled, the blast hurled everybody around like straw dolls, there was ear-splitting thunder. And the consequence? A gaping crater now stood in the place of the flat hill—the tunnel was surely unusable. But the already weakened wall had caved in as well, opening a breach through which a wagon could have passed.

In the morning Yao Yu-chung proposed to the Grand Council that the enemy catapults be destroyed by the fire drug. But his proposal fell on deaf ears. The first attempt to use *huo pao* in larger quantities had led to questionable success. Who could say what would happen next time?

In the afternoon the Tartars stormed Kaifeng with all their reserves. All the watchtowers were shot up in flames. Burning oil poured on the defenders of the wall from the raised platforms of the siege-machines. Fires broke out everywhere providing a flaming dramatic backdrop to the fall of the dynasty of the northern Sung. The Ch'in dynasty assumed the rulership of China for over a hundred years and, in turn, lost it to the Mongols united under Temujin.

Were weapons really employed in the struggle between the Ch'in and the Sung at the beginning of the twelfth century A.D. which we could describe as precursors of artillery? Before we turn to this question, we must first ascertain whether the basis of every fire arm, gun powder, also called black powder, was already known in China at this time. And here the answer clearly is yes. The aforementioned formula from 1040 A.D.

contains, along with unimportant admixtures, all the components of black powder. The mixture is also optimally approximated: 75% saltpeter, 15% charcoal, and 10% sulphur. The manual summed up the knowledge of that time. Therefore we are justified in assuming that powder had been invented before 1040 A.D.

The most important raw materials, saltpeter and sulphur, had been known in China long before then. Potassium nitrate (saltpeter) is found in Chinese soil in many places and is already mentioned as a medicine in the texts of the first century B.C. The Chi-yun Encyclopedia (605 A.D.) describes the way saltpeter is purified of undesirable components; sulphur is also described in detail. Saltpeter first appeared in the West in the thirteenth century as "Chinese snow."

It is no accident that chemistry developed relatively early in China. The Taoists believed that it was possible to find the elixir of life, the drug of immortality. Since the third century A.D. alchemists in China were very active and combined all kinds of substances. Now and then the Annals report on mysterious accidents, often connected with fire and explosions. Did one of them accidentally come upon the correct powder mixture? At all events saltpeter was considered so important in China that in 1067 A.D. an imperial decree prohibited its export.

The oldest accounts on explosions and fireworks that mark the morning of the Chinese New Year stem from the first century B.C. They deal with the sections of bamboo tubes which when thrown into fire burst with an ear-splitting noise. At the turn of the century, a new type of fireworks, the *pao chang,* achieved a great popularity. Judging from the description, it was filled with powder.

Huo pao, the fire drug, was also known early in China. With the exception of fireworks as such, however, the explosive and motive force of powder was disregarded. No written reports, at least, are known of, but we must bear in mind that strict security precautions and regulations also existed at that time!

In war the Chinese used black powder stirred into a paste in order to set targets aflame. The points of incendiary arrows were smeared with it as early as 969 A.D., hence two generations before the military manual appeared.

The first use of flame throwers is reported in 919 A.D. Petroleum, which is very hard to ignite, was set aflame by the fire drug. It was not until the eleventh century that the military leaders of the Sung dynasty harnessed the motive force of powder. They fastened narrow bamboo tubes onto the shafts of incendiary arrows and filled them with powder. The first step was taken on a long road that nine centuries later would arrive at skyscraper-high missiles of steel—an instructive example of the Chinese proverb: "Every journey of a thousand miles begins with one step."

There is a fundamental difficulty in the interpretation of Chinese texts concerning the origin of firearms. What is hidden behind the words listed in the Annals, Encyclopedias and depot lists? For example: "In the third year of the Hsien Ping (1000 A.D.), naval captain Tang Fu presented the Emperor models of huo-chien, huo-chiu, and huo-chi-li. He received a money gift."

These words are not explained in greater detail in the text. Only from the character *huo* that is common to all of them is it clear that it must involve the fire drug. Two specialists in the development of firearms in ancient China interpreted the aforementioned inventions of Tang Fu as incendiary arrow, incendiary bullet, incendiary bullet with hook (Wang Ling) and bomb, incendiary catapult, and hand grenade (L. Carrington Goodrich). The word "*pao*" is a further example of the problems of exegesis. Today it means cannon, yet originally it designated a sling-shot. *Pao* signifies the device that catapults projectiles coated with ignited powder. But *pao* is also the designation for the paper tube from which bullets were already fired. How are we to know now whether we are dealing with a primitive cannon or whether the reference is to a catapult?

In 1161 the fleet of the Mongolians was destroyed on the Yangtze Kiang by the fire drug. We are probably not wrong in assuming that powder was not employed as a motive force but only as an excellent incendiary device. Incendiary catapults,

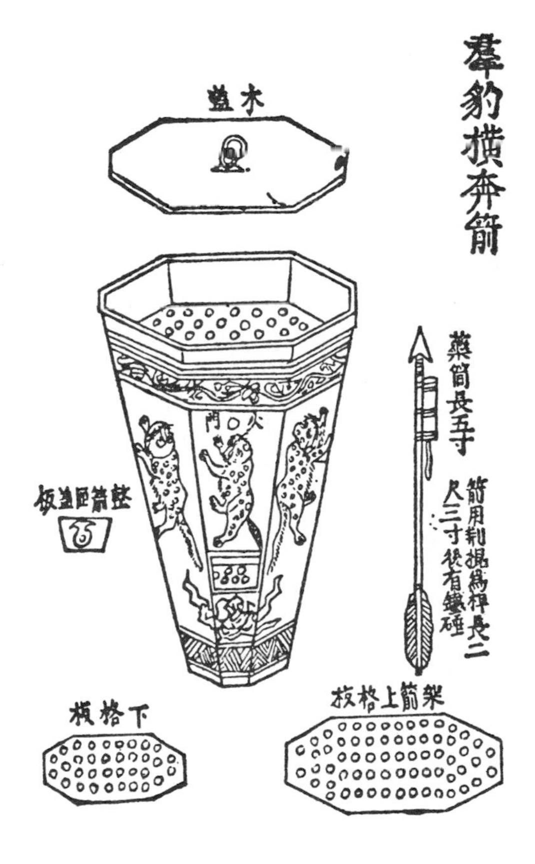

Chinese rocket arrow with starting device. From a military manual, 1621.

archers, and crossbowmen set the fleet of 600 ships afire. There is also an account from the year 1231 which clearly documents the use of powder as an explosive:

"During the battle a new type of artillery was employed, *chen tien lei,* heaven-shaking thunder. This weapon for the defense of the city, a catapult called heaven-shaking thunder, consisted of an iron container filled with the fire drug. The powder is ignited and the container catapults from it; the thunderous noise can be heard a hundred li away. The ground is churned up within the radius of nine acres, and wood and iron pierced.

"The Mongols worked at digging a narrow tunnel, through which only one man could pass, directly to the foot of the city wall. They shielded their operation with oxen skins. The Chin could not fight the Mongols from the walls. So someone suggested that the heaven-shaking thunder bomb, fastened to chains, be lowered over the tunnel. When the bomb exploded it tore oxen-skins and men into shreds, so that nothing of them remained." We can conclude, therefore, that the grenade preceded the incendiary missile.

In 1236 the Mongols invaded Kansu province. Governor Kuo Hsia-ma sent news to the capital that he was collecting all metals, including gold and silver, in his sphere of influence in order to cast *pao* out of them. Did he mean cannon barrels? We do not know.

Not until twenty years later do we unmistakably find firearms on the depot lists of the Imperial Arsenal of Yangchow: stones were shot out of strengthened bamboo tubes. Bamboo is a rather suitable material for a cannon barrel; bound with strong cord it can withstand an explosive pressure of 6 atmospheres. Thus the first recognizable cannon goes back to 1259 A.D.

In this same period gunsmiths also made cannon barrels out of paper. Fashioned out of many layers of paper and fastened together with glue, their stability more or less matched that of the bamboo tubes. Metal casting technique was already widely developed in fifteenth-century China. Thus it is not surprising that iron cannon barrels soon replaced the bamboo constructions.

When the Yüan (Mongol) dynasty attacked Japan in 1274, "black dragons" and "fire tubes" accompanied the expedition. In 1275 these weapons were not only mentioned by official Chinese war correspondents but they also appeared in two independent Japanese accounts of the battles. The first reference to a cannon appears in 1275 A.D. exactly 51 years before mention of such a weapon in the West.

The expedition against Japan was a catastrophic fiasco. First of all a typhoon (the "wind of the gods") destroyed the greatest part of the expedition's fleet, then the Japanese shoguns forced the Mongolian troops to retreat. A further attempt to subjugate Japan also failed. The Great Khan were forced to realize that they could not extend their empire farther east, so they steered their armies westward, toward Europe.

Was it the Mongolians, perhaps, who brought firearms to China? Chronologically their emergence coincides with the first clear accounts of fire arms. All records, however, clearly point to the fact that the Mongols for a long time had employed Chinese cannoneers for their artillery. The fact that as late as 1280 the Yangchow powder magazine exploded sky high and destroyed the city because a Mongolian soldier had carelessly handled the fire drug suggests how little experience they had in such matters. Accounts of firelocks accumulated from the middle of the fourteenth century on. Before the Ming troops conquered Peking in 1368 and thereby destroyed the Mongolian dominion, cannon was used to weaken the walls of the Imperial City so that they could be stormed. Instructions on their use were committed to writing, and the production of cannon barrels soon amounted to more than a thousand pieces per year.

The oldest preserved cannon barrel was cast in bronze on March 11, 1332; it weighed 15½ pounds, was 14 inches long, and fired 10½ caliber shells. It is on display in the Historical Museum in Peking.

In contrast to the development in the West, firearms in China made very little progress, and then only slowly. The Chinese were terrified when in 1517 the first European ship, the Portuguese fleet of Fernão Perez, sailed into the harbor of Canton and fired their cannon in greeting. They had nothing even remotely comparable to oppose these weapons. The idea of offering resistance seemed absurd. The naval guns of the Europeans continued to be a trauma for centuries.

Whence did the Portuguese have these fabulous cannons? How had the development of powder and cannon proceeded in

Europe? The course of events repeated itself here according to the same pattern already met so often: The invention reached the West first at the end of the Middle Ages or was discovered centuries later than in China. Thereupon, of course, further development proceeded rapidly; within a few decades a lead was established that China could not catch up with for centuries. Such had been the case with the compass and the art of printing, and with the development of flying contraptions and firearms.

It is interesting to compare what happened in China and Europe when cannons appeared. In Europe the roar of the cannon announced the end of the feudal age. The individual knights could no longer go plundering with impunity and laugh derisively in their solid fortresses at the nominal ruler and the royal laws. The seats of the princes and the bishops no longer offered secure refuge. In the early Middle Ages insubordinate feudal lords had often enough degraded the kings and emperors to powerless figures. Now the bombardments were mounted in front of the once impregnable walls, and reduced them to rubble in a matter of days, thus breaking the power of the lords. When Charles VII of France conquered Normandy in 1449, his artillery traveled from one fortress to another, and none of them withstood the pounding for long. Sixty castles were stormed within one year–a feat that would have been inconceivable without cannons.

Gunpowder also put an end to the armored knight on horseback. Now the feudal lord was no longer superior to the citizen thanks to his armament and his good sword; the most ordinary man could shoot him out of his saddle. The self-consciousness of the cities grew. With the help of cannons, they could now defy the nobility. And who cast and sold the cannon and thereby became the gravedigger of the age of feudalism in Europe? The bourgeoisie.

It is not an exaggeration to assert that gunpowder led to the revolution in the European social order. What happened in China? Nothing. Powder and cannon found their place in the arsenal of the armies, but they never led to a complete

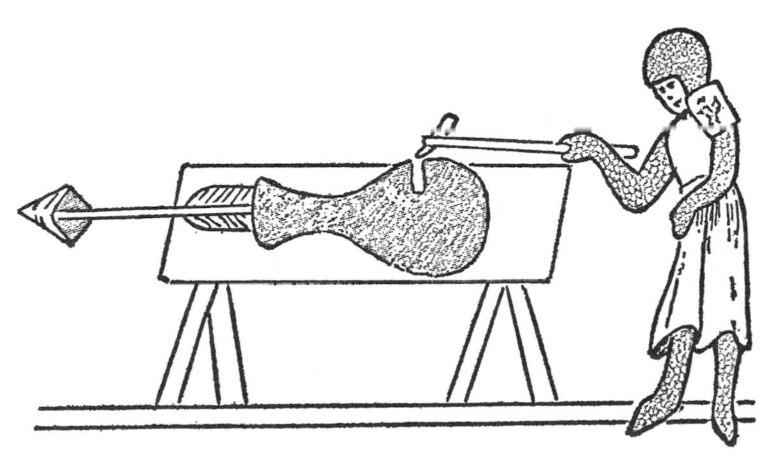

Probably the oldest picture known of a European cannon. Two manuscripts by Walter de Milimete (Holkham-Codex, 1326) supposedly contain illustrations of the new firelock. A knight holds a glowing iron to the touch-pan in order to fire off the bullet arrow.

subversion of the existing societal relations. Why? There was no system comparable to European feudalism. There were no fortresses and fortified places that the emperor had to destroy in order to assert his will. There were no local princes who could rule according to the motto: "Heaven is high, and the Emperor far away." At the turn of the millennium the central power was well established in China. The provinces were ruled by officials who could be replaced at any time, and who followed the directives from the capital city to the letter. Against whom should the new weapons of destruction be employed within the country? There was only one center of power, and that existed with or without cannons. Nor must we overlook the task that devolved upon the ruler of China: to make possible a life lived in virtue and dignity for all the people in his realm. This was an utterly different attitude than that prevailing in Europe, where man was exclusively an object from which the momentary ruler squeezed as much advantage as possible.

Inasmuch as the cannon in China was merely one weapon among many others, it was improved only slowly and Chinese ordnance fell hopelessly behind compared to the development that took place among the Europeans and Arabs. Inventive minds in Europe worked constantly to improve powder and cannon. Granulated black powder (1520) enclosed more air and thereby increased its efficiency. Ever larger ships carried ever heavier cannon. Landmines and shrapnel came into being. Time fuses were invented as early as 1585.

Black powder changed the world. Langdon W. Moore of New York figured out a special application: in 1866 he blew up the first Treasury with the help of black powder, came out of it with $200,000, and thus became the father of a profession still flourishing today.

12.

Silk, Caravans, and Profits: The Silk Road.

CH'UAN Sun sat down on one of the rocks that covered the shore on the left and right of the valley. In spring the melting snow in the Tien Shan mountains brought gigantic masses of water down into the river which transported sand, stone, and wood—always in great demand—to the steppes. Yet now, in the hot summer of the year 36 B.C., the valley had shriveled into a creek barely sufficient to serve as a watering place for animals. Somewhere farther north the river seeped away in the steppes.

For Ch'uan Sun they were well-known phenomena; as a chronicler—today we would call him a war correspondent—he accompanied the army of General Ch'en Tang on the punitive expedition against a tribe of the Huns. It was really more of an expedition than a campaign. The Han armies had long ago driven the predatory nomads into the steppes of Mongolia and westward through the Dzungarian Gate. Gone were the days when a Mao-tun could dictate his conditions to the Han. The giant empire of the Huns which only three generations before had been so powerful was once more split up into individual tribes, bitterly fighting each other. In the Imperial Court at

Ch'angan (in the proximity of the modern city, Sian), plans for further fragmenting the individual tribes were constantly being devised. At times the Han supported this or that group, at other times they bestowed their favor on another tribe. The aim was to prevent any grouping of the tribes in the north from again becoming so strong as to constitute a danger to the well-being of the population of China. To be sure the Han had further built up the extended fortresses and walls in the north and continued their construction even up to the Tarim desert basin. Nevertheless, the generals did not indulge in illusions: the defense of the Middle Kingdom did not lie in the Great Wall but in the strength of the Army. It was a component of the policy in the Imperial Court to bring the Ch'i-ch'i to reason and this was the purpose of Ch'en Tang's expedition.

It had been a long march; the soliders had endured great hardships on the northern edge of the Takla Makan desert. Although at this time the tributaries of the Tarim were not wholly dried up, they had had great difficulties with the water supply. But now they had arrived at their goal; the residence of the Ch'i-Ch'i lay on the other bank of the Talas and would not for long withstand the skill of the siege-engineers.

The city—actually it was an armed camp—was surrounded by an earth wall and an additional fortress, which Ts'uan Sun had not yet seen. A double row of pointed stakes guarded the city. In addition, the gates were defended by flanking towers made of thick beams. It was certainly an unusual defense installation in a landscape where wood had to be hauled from long distances. Something else had also aroused the professional interest of the chronicler, and that was the real reason why he was now sitting in the sun and spying, not without some eye-strain, across the river bed. A group of armed men, numbering about one hundred, was being drilled in front of the eastern gate. This was not especially surprising since the arrival of the Chinese armies had not remained secret and defense units could be seen training for the battle that was in the offing at all sections of the city wall. The movements of these soldiers were unusual: they repeatedly built a formation in which each shield interlocked with another so that a living wall was created. "Like fish scales," mused the observer. But then these scales suddenly separated, sword-blades flashed in the sun and the wall was formed again. The shields that covered the drilling soldiers were long and oval, wholly different from the round shields usually used by the inhabitants of the steppes. The distance was too great for the observer to distinguish their uniform or the facial features. It seemed, however, that the chieftain of the Ch'i-ch'i had taken into his service mercenaries from foreign countries. Ch'uan Sun betook himself back to the main camp. The general had to be informed about this. Who knew what other surprises the enemy might still have in store?

On the next day the siege-machines crept up to the walls, the soldiers stormed the palisades under the protection of undressed oxen-skins. The city fell. Among the prisoners were also the foreigners with the long, oval shields. Their battle formation had preserved them from great losses, and they had laid down their arms only because they had been hopelessly outnumbered by the attackers.

What does the chronicle from the Han time describe? Whom had Ch'uan Sun observed? Was it actually a group of Roman legionaries who had defended the city gate? This is how a number of famous scholars interpret the questionable text. That really would have been a remarkable encounter; not because they were so far from Rome–after all Roman troops had pushed as far as Scotland, southern Morocco, in the center of the Sahara, indeed, even as far as the upper course of the Nile, and the miles shrank in the march step of the cohorts. But it would have been the only direct contact between the two mightiest empires of the earth of this time: Rome and China. As the successors of the Ch'in, the Han had expanded the Middle Kingdom on all sides and for the first time had brought the Tarim basin with the surrounding mountains under control. The men of Han ruled from the Yellow Sea up to the Uzbekian steppes.

Farther in the West, Rome was rounding out its possessions.

It was to take only a few years for the Mediterranean to become a Roman lake, "Mare Nostrum."

Chinese expansion westward had reached its greatest extension. It would no longer make sense to go farther beyond the steppes or to scale the icy heights of the Pamirs. The westward expeditions by no means purposed to extend the borders of the Empire, and were limited exclusively to breaking the power of the traditional enemy, the Huns.

Rome, on the other hand, had deliberately extended its imperium farther eastward, but the resistance of the Parthians, who proved to be a formidable foe, was unbreakable. Thus in Asia Minor the Euphrates constituted the line of demarcation of the balance of forces. Barely twenty years before Crassus (nicknamed "The Rich") had to learn that the hard way: in 53 B.C. he lost the battle and his life in the struggle against the Parthians at Carrhae. Perhaps we can find the explanation for the presence of the mercenaries observed at Talas in this battle. Was it possible that the soldiers belonging to the legions of Crassus had entered into foreign service? Seventeen years and 1250 miles lay between Carrhae and the conquest of the city on the Talas. This would not be an impossible hypothesis. Besides this battle formation of the foreigners, which was utterly unknown to the region, there is another reference that hints at the presence of Roman legionaries. In the year 5 A.D. Chinese records make mention of a newly founded city in the province of Kansu (northwestern China) with the name Li-Kan. And this name speaks volumes. Li-Kan was the designation used by the chroniclers of the Han dynasty for the Roman Empire whose existence was known to the officials at the Imperial Court in distant Ch'angan through the Old Silk Road.

Why should a new city bear a name that designated a distant Empire? What is more obvious than the assumption that the mercenaries taken prisoner at Talas in 36 B.C. had withdrawn eastwards with the victorious expedition—perhaps even voluntarily—and settled down at a suitable spot? The Han generally did not slaughter prisoners, and Roman legionaries were good farmers and artisans as well as good soldiers. It would be logical for them to give the new colony a name whose sound recalled the homeland. A generation later the same locality received a new name which indicated that former prisoners had lived there.

This lost batch then mixed with the indigenous population. It would be a hopeless task to look for its traces now. Sixty to seventy generations have come and gone in the meanwhile. When the Nestor of China research, Professor Joseph Needham, in 1942, passed by in the proximity of the former city of Li-Kan, he believed to discern European traces in the features of the people living there. Yet it was clear to him that his imagination was playing a trick on him; he *wanted* to see these features.

Was this tiny western island in the sea of the Chinese population a one-time event? By no means. Around 1241 when Batu-Khan pushed toward Russia and East Europe, parts of his army swept through Silesia. Prisoners were taken, and for reasons that are still unknown to us today–perhaps they were exceptionally skillful miners–they were dragged through Poland, Russia, and the Asiatic steppes to the far side of the Balkhash Lake. There, on the bank of the I Li river in the region of the modern city of I Ming, they built their own settlement. The Silesians worked in the gold mines and as skilled manual workers they won the respect of the Mongols. This island also has left for us moderns no visible traces, and we would know nothing at all about this lost German enclave in Asia if Marco Polo had not incidentally reported it. When the Great Khan besieged an insubordinate city, the Polos served him as military advisers. The Silesians, who had been driven from their homeland, made the largest siege catapults whose shells broke the walls, according to their specifications. If these Germans had not enjoyed a reputation as excellent artisans, we would never have heard of them. For if Marco Polo had reported on all the Westerners whom he found in the service of the Great Khan, voluntarily or involuntarily, his book would have never been ended.

In point of fact it does seem that all contact between the

world empires in the East and in the West occurred through intermediaries. At all events there are no accounts that assert the contrary. Perhaps these accounts were never written or were lost or are lying somewhere in an obscure part of a distant library impatiently awaiting discovery. Certainly there was no lack of effort to establish this contact. Both empires were linked over long periods of time by the Silk Road and other trade routes. They could have both derived social benefit from direct contact. However, they were separated by many thousands of miles, and not only were climatic and geographical difficulties compounded, but a much more important impediment lay in the path of direct contact: the middleman. So long as nobody in Rome knew how cheaply silk was available in China, the middlemen could demand any price they pleased. The traditional foes of Rome, the Parthians, especially distinguished themselves in this connection. Their merchants derived such great wealth from the silk trade that its description by contemporary travelers simply could not be believed back home. Caravans had to pass through the Parthian empire if they wanted to reach the insatiable Roman market. And only Parthian caravans with Parthian merchants were allowed to take goods to the Bactrian border and transport them farther West.

The spice merchants traveled on the sea route through the Persian Gulf, around the Arab peninsula in the Red Sea. Yet that did not constitute an enormous detour for a traveler to Syria. Several hundred years later Chinese historians correctly commented on the abortive mission: it failed as a result of the profit mania of the Parthians. This episode produced a permanent effect: no Chinese ever reached a Roman province.

The Romans proved themselves to be presumably more successful in the pursuit of the same goal. Perhaps this was due to the fact that it was not an official who set out on the journey. Rather, agents of the Roman silk merchant Maes Titianus joined the caravans. Their purses must have been well-lined indeed since in the year 130 A.D. the Parthians were again defying Rome, and only enormous bribes could open the road to Roman traders. But Titianus had money enough. We do not know whether the travelers really came as far as China itself. It is a fact, however, that they did return from this journey and that from that time on the knowledge of the country of silk makers grew considerably. And this was also reflected in Ptolemy's world map. The otherwise so reliable Annals of the Han dynasty make no mention of the visit of the foreigners. Nevertheless, at this time when the Chinese Wall was opened, many travelers crossed the borders of the empire.

Long before the Silk Road united Orient and Occident—albeit only incompletely and with frequent interruptions—there was contact among the peoples of Eurasia. There are no written traditions dating from this time, but this is more than made up for by the language of the artifacts that the spades of archeologists bring to light. Let us consider the swords illustrated on page 178. Their form is typical for the Bronze Age in Western Europe, in the southern steppes of Russia, and in China. All four stemmed from the period that in Europe is connected with the Hallstatt civilization (ca. 750 B.C.). The conformity in shape of the blades will surprise no one; it was determined by the common function. But the unmistakable similarity of the handles cannot be explained thus. Here, quite clearly, a form was transmitted and, no doubt, from West to East, since these weapons appeared in China later than in Europe.

If the conformity among the bronze swords is amazing enough, it becomes downright bewildering with the axes. Here we are dealing less with axes for work or battle—the same function would compel a similar form—than with symbolical utensils, ceremonial axes. We can see that they are little suited for practical use. The edges are exclusively decorative, and the animal figures on the opposite side are ornamental. The similarity of these axes from China and the Hallstatt civilization, in no way necessitated by a common function, bespeaks an exchange directly across the Asiatic continent. Mo eover, these axe forms represent a considerable improvement over their predecessors. The handle is no longer fastened to the blade with bands and strings, but is socketed

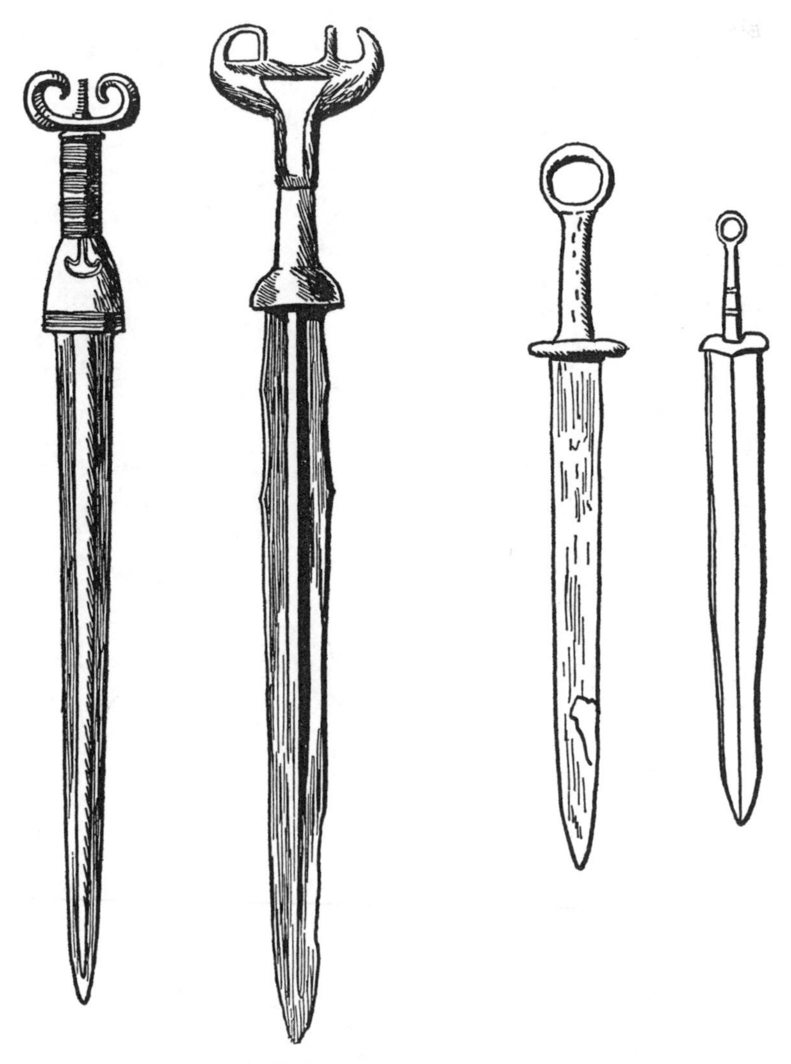

Bronze weapons from the Hallstatt Culture (Denmark and Russia) are depicted in the first and third drawings from the left. To the right of each is a contemporary weapon from China. Surely the similarity is not due to coincidence.

through a hole. Thus not only did this give the instrument greater stability and durability, but it also required much less of the precious bronze to make it.

Weapons and tools are not the only artifacts that point to a transasiatic exchange. Neck rings and other pieces of jewelry offer additional indications. Further, that the same image of a bird mounted on three wheels has appeared on bronze or pottery in Western Europe, in Egypt, and in China as well can be considered coincidence only with great difficulty. Finally, the motif of "the flying gallop"—a horse thrusting forward at an extremely high speed—from 1000 B.C. on spread throughout the Mediterranean world and across the South Russian lowlands up to China. Its appearance over such a wide area cannot be attributed to the observation and representation of this movement by the horsemen of the steppes, for no horse gallops in such a manner, as a slow-motion film will easily show.

Thus there are many mute attestations that in the Bronze Age a lively reciprocal action between the two ends of the Eurasian continent took place. And why, for that matter, should this not have been the case? An infinite expanse of steppes runs from the Hungarian lowlands up to Dzungaria at the foot of the Altai mountains. It was an excellent migration route for pastoral and hunting tribes. There were no political

This grave relief from the time of the Han Dynasties shows a position that has no counterpart in nature: the "flying gallop." The motif was brought to China from the Mediterranean region.

frontiers at this time, at all events none that would have been a serious obstacle. The exchange must have taken place across the steppes belt which at that time was the path of least resistance. The route across the broader deserts lying southward and across the high mountain-chain—preferred later—already presupposed a certain organization of the caravan routes.

As is so often the case, war once more was the father of all things. Along with the rule over the Middle Kingdom, the Han

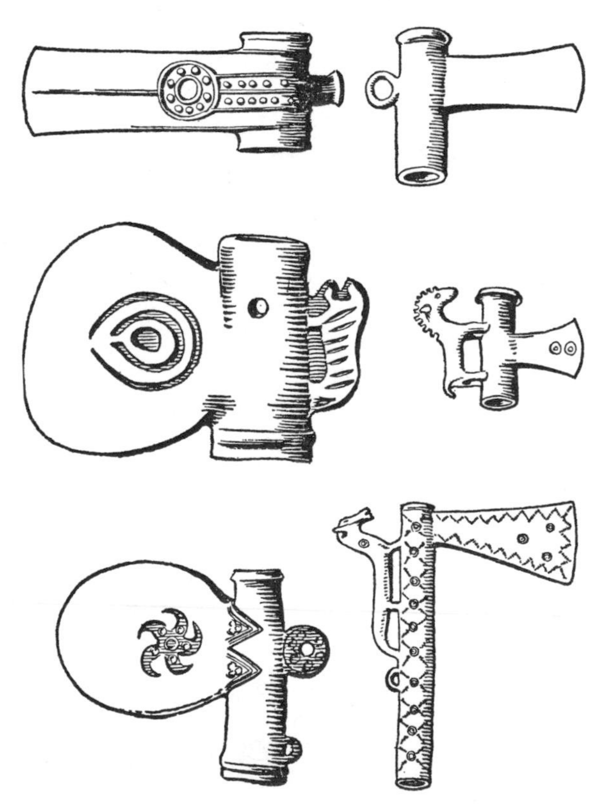

Ceremonial axes from the Bronze Age. Left: axes from China. Right: finds from the period of the Hallstatt Culture in Europe, ca. 750-450 A.D.

dynasty in 202 B.C. had also assumed the heavy burden of protecting the northern and western borders against the Huns. These nomads, ever bent on plunder expeditions, saw in the fertile valleys of China what the Vikings glimpsed in Normandy a thousand years later: rich expanses of land which lay open to anyone who reached out for them with sufficient zest and vigor. In the beginning, just like the Vikings, the idea of settling in these regions was the farthest thing from their minds. A swift plundering expedition and then back to the native hearth was the plan of action.

In the beginning of their rule the Han were still too weak to resist the attack of the Mongolian cavalry formations. They deflected the thrust of the Mongol tribes, united under Mao-tun, westward while they sent endless caravans to the North, laden with tribute payments. But they did not fail to take advantage of the respite. The central government systematically built up its power. The irrigation system, the building of which had begun under the Ch'in, was now completed and assured regular and abundant harvests. Thus a part of the population was freed for military service and soon the Han armies marched along the roads radiating from the capital Ch'angan.

When Wu Ti ascended the Dragon Throne in 140 B.C.—he was to rule for 54 years—the time was ripe finally to liquidate once and for all the threat posed by the tribes roaming the northern steppes. At first the Great Wall, stretching from the Yellow Sea up to the Nan Shan mountains, was extended farther. But, as experience taught, that was only a weak defense against the highly mobile nomads. Wu Ti planned to seize the Huns within the prongs of a strategic out-flanking movement. The Tokharian people were China's natural ally. They had lived in the steppes and oases of west Kansu since time immemorial until the Huns came and pushed them westward. It had been more like a panic flight than an orderly withdrawal. Terror pressed at the heels of the entire people and they had only one aim: to establish as much distance as possible between themselves and the Huns.

Wu Ti attentively read the statements made by captured

Huns which the garrison commander in the northwestern part of the country had dispatched to him: "The Hsuing-nu [Huns] have overcome the king of the Yueh-chih [Tokharians] and fashioned a goblet out of his skull. The Yueh-chih have left their country and are hiding somewhere. They are imbued with a desire for revenge, but they have no allies with whose help they could beat the Hsiung-nu."

"That can be remedied," said Wu Ti to himself. Thereupon he summoned his highest officials. The servants of the State were soon agreed among themselves: contact should be established with the Tokharians and then both should attack the Huns at the same time. But where were the Tokharians? Someone had to journey to their redoubt as ambassador. This was a difficult and dangerous task because the Huns controlled all roads in the north and west. Whoever wanted to reach the Tokharians had to go through Hun-ruled territory. Chang Ch'ien volunteered for the mission that appeared to border on the impossible. He started out in 138 B.C. and promptly fell into the hands of the arch foe as a prisoner. He was forced to spend ten years in their camp. But it was a comfortable stay in that he neither went in chains nor suffered hunger or any other tribulations. Nevertheless, all the servants and the large tent at his disposal could not dispel the realization that he was indeed a prisoner. No guards stood in front of his *jurte* because nobody could escape the Huns in the steppes unless he was a far better horseman than they. Ten years went by before Chang Ch'ien came into the possession of a superb horse and an experienced guide. He had, no doubt, also blinded his alert-eyed persecutors with gold; otherwise, he would have never been able to make a successful flight to Ferghana. There, in the Uzbekian steppes the horse-breeding nomads could point out to him the route to the Tokharians.

The Tokharians had in the meantime overrun the Hellenistic state of Bactria and settled down in its fertile valleys and rich meadows. Ferghana had every reason to rejoice over these invaders because now its inhabitants were free of the rule of the Bactrians. Chang Ch'ien drove farther south with his entourage to the camp of the prince of the Tokharians. The envoy of the Han was received with open arms; all conceivable honors were showered upon him, all his wishes anticipated and instantly gratified. The king was wholly deaf to only one thing: when the conversation turned to the subject of a military expedition against the Huns, he shook his head. No, never before had his people had it so good. Now they were sitting by the fleshpots, they were the lords of Bactria and never even dreamed of jeopardizing their new-found affluence for the sake of a long-forgotten desire for revenge. Every wish of the men of Han would be fulfilled. The Tokharians would even send a counterpart mission laden with gifts to China, but a war against the Huns was absolutely out of the question.

Chang Ch'ien had to content himself with that. He remained with the Tokharians for one year and utilized this time to gather all available information on neighboring countries. The guides of the trade caravans coming from far away always found him a willing listener for their tales and one whose cash box was always dipped into when they brought him rare plants, animal or farm products. It was only for expensive silk that he had no use; he was interested merely in knowing where it came from. Judging from the samples the silk stemmed from the province of Szechwan. Fine. But along which route did the bales arrive in Bactria? And it was then that he learned about the trade route across the heights of Yunhan through the primeval forests of Indochina to India proper. From there the caravans brought the coveted cloth to Bactria and farther westward to the kingdom of the Parthians. Chang Ch'ien cocked an attentive ear.

He also committed to writing many other reports. Among them there was one about an expedition which the Bactrians had carried out more than two generations before. The lucrative flow of gold from the countries lying northeast of Bactria had suddenly ceased, so a selected group of soldiers and merchants was sent out in order to determine the cause. Their mission was a failure because blue-eyed, red-haired men repulsed their advance. Chang Ch'ien could pinpoint where, and explain why, the flow had been interrupted: the Huns cut off trade with Siberia. Yet he did not know, nor do we, the identity of the people who had forced the Bactrian expedition to retreat with battered and bloody heads.

Chang Ch'ien set out for home again in the year 127 B.C. Although he carefully chose a wide detour, he fell into the hands of the Huns again. But this time they held him prisoner for only one year. Once again it was just a kind of honorary house arrest; none of his possessions or notebooks were taken away from him or lost. In 126 B.C. he finally passed through the gate of the capital city of Ch'angan. One of the longest journeys of antiquity had come to a close. He could not present the Emperor with any kind of a military alliance with the Tokharians. Nevertheless, Wu Ti viewed the mission as a smashing success, and rewarded Chang Ch'ien with a high government office.

The Emperor's advisers immediately sifted the findings of the twelve-year journey: now at first hand there were reports concerning the topography and the kingdom of Ferghana, Bukhara, Samarkand, Tashkent, and Bactria lying in the far west. Reliable reports concerning the steppes south of the Sea of Aral, the countries north and east of the Caspian Sea, the empire of the Parthians, Mesopotamia, and India were also found among the records of the expedition. And these were by no means only geographical tractates: strategic localities, foreign techniques of warfare, the potentialities for economic expansion and for establishing political relations did not go unmentioned. In particular Chang Ch'ien repeatedly stated that nowhere had he heard anything about sericulture, hence silk was a non-competitive article of trade.

The mission brought back the grapevine and alpha grass with it from the countries beyond the Pamirs. Later these were followed by figs, cucumbers, chives, sesame, pomegranates, and walnuts.

After all aspects of the long journey had been fundamentally appraised, Wu Ti summed up the policy that he planned to pursue in northern and western China in the years ahead:

"Ferghana, Bactria, and Parthia are great kingdoms full of precious things and inhabited by a sessile population that pursues activities similar to ours. But they have armies of no consequence and they highly appreciate China's products. In the north live the tribes of the Huns, war-tempered peoples with whom presumably we can best promote our interests with gifts and bribes."

Wu Ti wanted to stabilize the northeastern borders of the empire. He had two means of achieving this, the "carrot and the stick": silk–and soldiers. He went to work resolutely. It was necessary first of all to establish the connection with Bactria and to lay out a trade route. The direct way was closed off by the Huns but neither did it appear practically suitable on account of the vast deserts and the chain of mountains that could be negotiated only with the greatest difficulties and hardships. Chang Ch'ien had taken special pains to point out that travelers at these great heights would be exposed to all kinds of inconveniences such as nausea, headaches, and nose-bleeds. A new mission tried to reach Bactria along a route running deep in the south. It was never heard from again and presumably it had fallen into the hands of the wild tribes of Burma. The only remaining alternative, therefore, was to fight to free the direct route to Bactria.

The Han armies achieved this aim in 121 B.C. Chinese influence now extended up to Lop Nor, and the whole Tarim basin, up to the heights of the Pamirs, lay open to the grip of the Chinese armies. The direct route was free!

Workers followed the soldiers, the defense installations of the Great Wall soon reached as far as the salt pits on the western slope of the Nan Shan mountains. Again the Great Wall served more as a symbol of the territory under Chinese rule rather than specific military purposes. Where the towers and walls were raised, there ran the demarcation line between the civilized world of China and the realms of the barbarians. This did not necessarily mean that the Chinese did not want to have anything to do with these peoples. On the contrary, at the time of the Han dynasty the gates of the Chinese Wall were wide open and no month went by that did not see at least one brilliant embassy, laden with gifts for the "barbarians," pass through the gates. They betook themselves to all the peoples of the West in order to bear witness to the greatness, the might, and the glory of China. How it flattered the Emperor of the Dragon Throne when he heard that the delegation destined for Parthia had been met at the border by an honor guard of 20,000 horsemen.

It could not but rebound to the prestige of the Han when they received counterpart envoys from distant lands and extended to them the most cordial hospitality. Thus nothing was spared when an embassy of the "barbarian peoples" paid a visit to the Imperial Court. Just as in the twentieth century corporations entertain potential customers, so did Wu Ti offer all the treasures of his kingdom to foreign dignitaries in order to impress them. It was presumed that they would then proclaim to the astonished world outside precisely where the true center of the earth was to be found. The visitors then returned to their homeland richly laden with gifts and with their senses still reeling because of the splendor which they had beheld. But Wu Ti was not concerned only with the prestige of the dynasty in the world beyond the Wall. The embassies from distant lands also buttressed the power of Wu Ti in his own country. Their gifts were "refunctioned" into payments of tribute. The influence of the old feudal lords had not wholly disappeared, the hierarchy of officials did not yet have total control of all domestic affairs. But the establishment of the absolutist bureaucratic state was Wu Ti's declared aim; and the higher the emperor's prestige rose, the more the influence of the local lords was bound to disappear.

To be sure, this entailed enormous expenditures for the Imperial Court, since each diplomatic mission consisted of more than a hundred officials, not to mention the gifts and the baggage train. But accounts were squared in the end. Glass, imitation precious stones, jade, fine wool and linen fabrics, horses, rare plants, magicians, artists, and much more that was much in demand in China poured into the Middle Kingdom as gifts in return.

The imperial envoys were soon followed by the transports of merchants who made no secret of their intention of earning handsome profits in the exchange of goods. And they brought with them an article that met no competition and whose price they could fix almost at will: silk. But the preliminaries were always the same: the Emperor's envoys were the first to set out and soon thereafter traders followed in their footsteps.

But the carrot was not always willingly accepted, and when this happened the stick was in ready reserve. Ferghana's fate was still clearly etched on the memory of all of China's trading partners: they had received the Emperor's gifts with pleasure but had lent a deaf ear when the envoy had hinted that his lord expected some of their superb "blood-sweating" horses in return. No, they did not want to be separated from their splendid steeds. The envoy then became brutally frank, impolite, insulting–they showed him the door. They did something else–heaven was high and the Chinese Emperor 2500 miles away–they lay in wait for the caravan beyond the country's borders and beat all the members of the delegation to death. The garrison in the Great Wall nearest to Ferghana was rushed to that country to bring punitive action but suffered the same fate.

Two thousand years later the commanding generals in the Pentagon were to make a hobby out of the domino theory: if Vietnam falls, Laos falls; if Cambodia falls, Thailand falls; if Thailand falls–and so on. Wu Ti surely is unknown to these gentlemen, yet they are merely repeating his arguments: if Ferghana can commit this with impunity, Bactria will be quick of hearing, Parthia will follow suit and the tribes of the Hun will not tarry. Wu Ti, of course, had one advantage over his later imitators: he could prevent the first domino from falling. His order set 60,000 soldiers, 30,000 horses, 100,000 oxen, and an endless baggage train on the march. These masses of men and animals had to be guided through the Tarim Basin along different routes, otherwise they would have drunk dry the rivers and streams en route. The army wound its way like an enormous worm to Ferghana between the Pamirs and the Tian Shan mountains. The Khans did not realize that they were not being butchered just so that the Chinese could get their hands on their fiery steeds. At stake now was China's prestige and influence as a great power. She could not afford "to lose face." The dream of an independent Ferghana was shattered in 114 B.C. The only remaining alternative was flight along the shore of the Jaxartes river in the steppes lying farther north. The price for the "blood-sweating" horses dropped drastically.

After the matter was settled to his satisfaction, Wu Ti sent ten more embassies in the direction of the West. They were equipped even more magnificently than their predecessors, they still showered kings, princes, khans, and chieftains with precious gifts—silk above all. But was it possible that a soft threat vibrated in their well-phrased utterances, that the word "Ferghana" hovered invisibly in the air when they spread out their gifts? Be that as it may, the Chinese envoys were received everywhere with respect and dismissed with princely gifts in return and promises to visit the Emperor in his residence as soon as possible. Silk merchants followed the diplomats. Twenty years had gone by since the end of Chang Ch'ien's journey, meanwhile become legendary, and the arrival of the first trade caravans, setting out from China, in the Parthian kingdom. The Silk Road began to throb with activity.

How quickly the ladies of Parthia grew accustomed to the luxurious silk. It brought a whole new feeling to life. The possession of silk began to seem a vital necessity; once one grew used to it the thought that one would have to forego it was intolerable. No matter what the scheming politicians were concocting, the prosperity of the silk trade always had to be taken into consideration. Wu Ti, who still wore the yellow robe of the ruler, could be satisfied. Trade had stabilized his northwestern borders. Also, within the country no one questioned his authority or that of the bureaucratic State.

Silk from China flowed in a never-ending stream by way of Turkistan and Bactria into the kingdom of the Parthians. The

Parthians were learning that silk was not only a splendid item of clothing but would also be a valuable commodity for trade. Farther west there was an insatiable market that would absorb any quantity at almost any price. The Roman Empire had incorporated every adjoining state along the Mediterranean, one after the other. Since the Romans had conquered Crete in 67 B.C., their dominion now also spread over the east end of the Mediterranean, and they had no trouble in making short shrift of the pirates operating there. The silk transports could arrive unmolested by ship to Rome, instead of along the tedious land route. This was an important factor in the silk trade.

What the Roman ladies wore instantly became the fashion in the provinces, from western Spain to Syria, from Gaul to Cyrenaica. And what material was better suited to show off the beauty of a Roman woman than silk? Their husbands, suitors, and lovers would pay any price for it. The Parthians knew this and acted accordingly. At this point the naive may object that the Parthians were involved in an ongoing struggle with expanding Rome and would not engage in trade with the enemy. Well, of course, war is one thing, trade another. and the traders at any time could trump objectors with the patriotic argument that the enemy was being hurt when by being sold luxury goods at high prices. The sesterces spent for silk would not be available for the purchase of weapons and for paying the legionaires! Nevertheless, appearances were preserved. The trade between the Roman province of Syria and Parthia was conducted through buffer states that were allowed to exist in reciprocal understanding.

But here we are talking constantly of producers, middlemen, and consumers and we completely forget the blood of this trade artery, whose existence kept this remarkable contact between the Far East and the West alive: the silkworm. An unpretentious moth *(Bombyx mori)* kept the Silk Road going. If we are to believe legend, a goddess had taught the art of sericulture to the inhabitants of the plains of the Yellow

River–the region of the present-day province of Shantung–4500 years ago. But those with a penchant for facts should be told that silk is mentioned for the first time in 1240 B.C. in the reign of the Shang dynasty. By then dwellers along the Hwang Ho had already mastered the art of silk making with a perfection that could only be the result of long experience. This is confirmed by silk residues found in the tombs of the Shang period.

We owe the reputation of Chinese silk from earliest antiquity to the present day in good measure to the governmental authorities who scrupulously supervised the quality of the raw material and of the processed fabric. This quality control has been maintained over the centuries so that today it is still impossible to sell inferior silk from a Chinese province. This does not mean that all silk is alike–each silk-producing province concentrated on a specialized product–but the official seal guarantees that the quality achieved in the past has not changed. This authority, already institutionalized under the Han, is the pioneer of modern-day government quality control agencies.

The question of how much silk was produced in ancient China should be considered at least briefly. First of all, some figures. About 1000 silkworms are hatched from 15 grains of eggs. They eat about 65 pounds of mulberry tree leaves until pupation. About six ounces of silk are reeled from these cocoons. In addition we must take into account that the path from the egg to the onset of silk spinning up to the complete cocoon is a long and tedious process. Thus it is difficult to answer the question of how much was actually produced. Here we are forced to rely on indirect references, for example on payments of tribute.

When the Han dynasty came to power around 200 B.C., the empire was extraordinarily weakened because of the struggles over succession to the throne. The nomadic peoples in northern China had joined the great empire of the Huns under Mao-tun. The Han had hardly any wherewithal with which to oppose the onslaught of the Mongolian horsemen. Mao-tun could have

Silk is reeled from the cocoon.(Chinese drawing, 1637.)

conquered China down to the Gulf of Tonkin without meeting serious resistance. China bought her freedom. With what? With silk—along with other treasures and the so coveted Chinese princesses. Several hundreds of thousands of rolls of silk persuaded Mao-tun to seek his happiness farther west. For Europe this was a decision of far reaching consequence; the silk tributes deflected the conquerer's assault westward, the assault wave spread to the Western world and triggered the migration of peoples. It was silk, in enormous quantities, that saved China, not the Great Wall.

In the course of history China repeatedly purchased her freedom by means of silk tributes, at least for a while. From these deliveries we can infer the vast amounts of silk that were produced. Nor were supplies at all exhausted after payment of the reparations. In order to view these figures in the proper light, we must remember that the population of China was estimated to be around 50 million at the time of the Han dynasty (202 B.C. to 220 A.D.).

Silk rolls first and foremost made the journey from the East to their distant markets in the West in jolting ox carts and on the swaying backs of donkeys, camels, or dromedaries. Along this road, exposed to the perils of all kinds, it was worthwhile to transport only wares which were precious and which, though of high value, were light and required little space. The merchants who left the Jade Gate in China were not the same ones who passed through the Syrian deserts after many months and who unloaded the precious bales in the bazaars of Damascus, Sidon, Aleppo, or Gaza. Silk was a transit commodity and the merchants took advantage of every opportunity to trade in additional local wares. Thus the pack animals were heavily laden with skins, wools, rugs, and agricultural products—in short, with everything that could bring in some money in the commerce from border to border. Local traders also always joined the silk caravans. They traveled with relative security, since the transports were always under armed escort. The aim of many robbers or local potentates (the distinction often enough was blurred) was to plunder a silk caravan. Hence constant alertness was advisable. Not seldom a caravan of this kind included ten thousand draught animals and a like number of merchants and soldiers.

Although silk dominated the business dealings from the East to the West, the caravans from the opposite direction carried a significantly greater variety of goods. They were forced to do so since the West had no single product that could even approach silk in importance. In addition, China was richly endowed with the fruits of the earth and at the time of the flowering of the Silk Road was technologically considerably ahead of the trade partner on the other end. Hence it was no easy task to find goods that could be of interest to China. But there is no need to speculate as to what they were, since the depot lists of the Han provide us with exhaustive information. Glass from Alexandria and Sidon is prominently mentioned. It was highly valued at the time of the Han since China was dependent on natural, transparant crystal. It was not until the fifth century that the artisans of the Middle Kingdom also began to melt glass. Of the glass wares the Imperial Court was especially smitten by imitation precious stones, glass goblets, containers, and vases always found ready buyers. Incense and purple dye were found in the bundles alongside amber, which had already traveled from the Baltic Sea coast to Rome. Artists prized imported jade since the qualities of native jade did not meet their requirements. Fine wool and linen fabrics were coveted in the land of silk, as were gossamer-thin fabrics made by the artisans of Li-Kan, as Seleucia Media, Syria and Egypt were known. Hardly any of the admirers knew that the raw material for it–silk–had passed through the Chinese border in an opposite direction only a few years before.

Yet all these goods did not suffice to pay for the silk imported by Rome. The balance ensued with precious metals. The merchants were especially interested in silver. Gold in China frequently had only a metallic value, but not a nominal value. Gold and silver probably arrived in China in the form of bars, since there are hardly any places where Roman coins have been discovered. In contrast, archeologists in India–which served as middleman in the maritime commerce in silk and spices—have discovered countless coins from the Roman period. We shall discuss the difficulties of making up the adverse Roman balance of payments later.

No depot lists receipts or reports from embassies have witnessed the thoughts, ideas, conventions, and conceptions that traveled with the wares along the long dusty road. These were light, yet nonperishable commodities, whose traces we can seldom follow. This is relatively simple in the case of the new religions. Thus did Buddhism and Manicheanism travel along the stations of the Silk Road in the direction of China. But what happened with secular ideas? What did the Tocharian merchants relate to those from Parthia around the evening campfire? When they, perhaps, had heard of an invention—at second hand—and spread the news further, what a source of misinterpretation and mistakes they became! No doubt most of the ideas, half undertood, embroidered with fancy, and mutilated by translators, never got past the road.

Nevertheless there seem to be several examples of the transmission of intact ideas along the Silk Road. It is probable that geographical knowledge was transmitted orally. Strabo (63 B.C.-20 A.D.), for example, knew practically nothing concrete about the empire of the Chinese. On the other hand Ptolemy (85-160 A.D.) could already draw a map on which China was represented. It is particularly instructive that Lop Nor, a wandering lake not far from the Silk Road, is to be found on the map which otherwise represents few details. No doubt Ptolemy's informants were very familiar with the Silk Road.

An even more convincing example of the exchange of non-material goods along the Asiatic trade route is the following: In the Byzantine empire of the fifth century and in the subsequent Islamic empire there were official clocks that struck the hours through a mechanism that caused a brass ball to fall into a bronze basin. Flaps that lifted to reveal figures or lights indicated the time of day or night. These clocks were driven by water power. The visitor to Morocco can still view the sorry remains of this former splendor today in the Medina of Fez opposite the Madrasal (a Moslem mosque school) Bu'Amaniya. These clockworks are described as treasures possessed by the people of the West in Chinese texts from the tenth century. This knowledge must have wandered from station to station along the Silk Road.

We could speculate for pages on end on the transmission of ideas, conceptions, and–above all–inventions along the Silk Road. Yet prudence is advisable because one can all too easily arrive at false conclusions. This is what befell many researchers in the eighteenth century when a stream of Chinese philosophy, art, and science flowed to Europe. All the inventions of mankind were said to have been cradled in China and to have radiated in all directions by way of the Silk Road. This is one extreme, yet we must also avoid the other which was so modern at the beginning of our century. Since then, China has been considered merely a recipient of ideas without having made any contribution of her own to the great ideas of mankind. Only two things are certain: an exchange of ideas between the Western world and the Middle Kingdom took place over the Silk Road. This exhange, however, was not intense enough to leave traces on one side or the other.

Up to now we have interested ourselves only in the particular motives and the particular historical background from which the classic Silk Road developed. Many ideas and products that traveled with the caravans from East to West and from West to East have also been touched upon. But the reader has still learned little about the actual route. And this was precisely what interested the traveling merchant most. What did he care about the political interconnections, as long as border disputes did not lead to the closure of the road and the tolls and tariffs of the individual overlords did not soar to unreasonable heights? He wanted to know what it would cost him along the journey since preparations for it depended precisely on this intelligence. And he could then stow away goods in order to pocket an extra profit on the side.

Let us imagine a bale of silk that stretches from China along the Silk Road, in its whole length until it arrives in Syria. Let us follow its path, but let us not forget that most of the merchants to whom the bales brought profit turned back at the next land border. At the time of the classic Silk Road, from around 100 B.C. up to the end of the eighth century, it is probable that no trader made the whole trip along the entire road with the caravans.

One problem emerges: where did the Silk Road begin in China? As a choice we have Loyang, the old Sinae Metropolis and Ch'angan (Sian) that was named Sera Metropolis by the geographers of antiquity. Both places bear the surname capital. And rightly so, since Sera Metropolis was the capital of the earlier Han dynasty and Simar Metropolis that of the later Han dynasty. It was believed, even up to the seventeenth century, that they belonged to different kingdoms. No wonder, then that in the Middle Ages the missionaries, travelers, diplomats and merchants reached Loyang by the sea route around Indochina and then continued their voyage upstream on the Yellow River. The foreigners saw Ch'angan after many months of a protracted journey that led exclusively over a land route. Geographical bearings could not yet be taken with sufficient exactness with the instruments of that time so that the geographical position of both kingdoms could altogether have been very different. Nevertheless the dispute over the beginning point is pointless. Nowhere did a sign point to "Silk Road, X miles."

The silk in Ch'angan waiting for transport already had a long journey behind it. Most probably it stemmed from the cultures on the lower course of the Yellow River, but also from the province of Szechwan and the region around Canton supplied the merchants. When they left Ch'angan the silk caravans plodded along in a northwesterly direction through Lanchow and Suchow parallel to the chain of the Nan Shan mountains. The routes presented only few difficulties to men and pack and draught animals but other perils lay in wait: Tungusic bandits threatened this natural postern gate of China from the south, while in the north the Huns, ever bent on plunder raids, repeatedly sallied forth to attack the caravans.

Further west the caravans passed the Jade Gate, the end point of the main line of the Great Wall; a relatively weak section still stretched considerably further West almost up to the swamps of the eastern Tarim basin. Here was the gate that separated the Middle Kingdom from the barbarians. Homesick Chinese named it the "Dragon Mouth": it spewed forth the traveler. Those who passed through always remembered to lift a stone before the gate and throw it against the wall. If it fell flatly to the ground with a dull thud, the heart sank because it signified that one would not return. On the other hand, if it rebounded with a ringing sound, the homesick traveler quickened his pace in order to shorten the time between his departure and the promised return. Not everybody passed through the Jade Gate voluntarily. Along this way many a princess had reached the distant place where the prince's son—to whom she had been allotted as his consort for political reasons—was waiting for her. At the order of the emperor the Jade Gate also spewed forth officials guilty of serious malfeasance in office. Only seldom were they permitted to return, yet they always made preparations at least to be buried in the good earth of China.

Here, too, the Silk Road forked. Not without reason: a natural obstacle closed off the route to the West, the Tarim basin with the Takla Makan desert. It is one of the most arid regions of the earth and practically impassable, at all events for any large transport caravans. On his historic mission Chang Ch'ien had gone along the southern end of the desert, a route that leads through Chotan to Kashgar. Marco Polo probably also rode along here in the thirteenth century.

Since it was not advisable to take the southern route because of its extreme aridity or the bandit attacks of the Tibetans, a caravan chose an alternative which had been increasingly resorted since 5 A.D.: from Tunhuang along the shortest way to Turfansenke. In this fertile group of oases the caravans halted to rest and gather new energy and then pushed on further, either through the desert on the left, or across the heights of the T'ien Shan mountains. In Kashgar the trade route linked up with the route that goes around the Tarim basin on the south side.

For a while a third route was available. It led to Lop Nor, followed the oasis chain of the Tarim River and then arrived at the foot of the T'ien Shan mountains on the alternative route. The exact course can no longer be reconstructed today, since the Lop Nor (Lop Lake) has strayed strangely in the true sense of the word. Sven Hedin rightly called it the "Wandering Lake." Its existence was known for a long time (it is the northeasternmost lake on Ptolemy's world map), but it was not until 1877 that a Russian explorer established its exact position. The Tarim River ends in it; it can be described more accurately as an enormous swamp than as an open body of water. Whenever the Tarim changes its course, the Lop Nor also "wanders." It has changed its position often within the last two thousand years, and drastically so: the newborn lake could be as much as 185 miles distant from its former dried out bed. It is no wonder that it had been a puzzle for geographers for so long, which was first solved by Sven Hedin.

Around the time of Christ's birth its position was suitable for crossing the Tarim basin, and an important caravan base arose on its shore. In the fourth century water from the Tarim no longer reached this location; the lake dried up, the city vanished, and caravans no longer came. No water meant no life. It was not until the middle of the twentieth century that the region aound Lop Nor awakened to new life. The lake was recalled with a bang: it was here that the Peoples Republic of China exploded its first atom bomb on October 16, 1964.

Most of the time the border traffic with Parthia proceeded smoothly, because the Parthians too had a good head for business and had no interest in jeopardizing the continuity of the flow of goods. They had established themselves there at just the right time. Around 250 B.C. north Persian tribes conquered the satrapy of Parthia, extending their rule up to the Euphrates. There Parthia bordered on Syria and resisted the expansionist drive of the Roman Empire. The Parthians were superb warriors who inflicted one defeat after the other on the Roman legions. We must confess, too, that they were even better merchants than soldiers. They immediately recognized the importance of the trade link with China, since the envoy of Emperor Wu Ti was greeted with unusual honors at the border and escorted to the capital. The Parthian merchants must have rubbed their hands gleefully when the first through caravan coming from China reached Parthia in 106 B.C. They were sitting on the most important section of the Silk Road; nobody could pass them unshorn, since all routes led through Parthia. In the north the Caspian Sea closed the route to the West and in the south traders did not exactly find the turbulent waters of the Persian Gulf inviting. The Parthians had to behave very unreasonably before the silk merchants would change to alternative routes. In point of fact they did later push their advantage to such an extreme that they practically ruined the silk trade for all time–and themselves as well. Actually the Parthians themselves were too shrewd to do a thing like this. In point of fact it was the Sassamids who conquered Parthia in 224 A.D. and founded the second Persian Empire, who turned the screws so hard that it killed the victim in the process.

The Parthians not only controlled and secured the course of the Silk Road from Merv to Mesopotamia, but they also had an excellent knowledge of the market for goods produced in the Orient. Only they knew what price was paid for silk in the Roman Empire, and also that, practically since the time of Augustus (63 B.C.-14 A.D., emperor of Rome 31 B.C.) Romans would pay any price for it. They owed their measureless wealth to this knowledge and to their monopoly in the market. Astonished contemporary travelers described the gold- and silver-plated columns of the villas of the merchants, who defended this privileged position with all the means at their disposal. An impressive example of how they set about this without raising premature suspicion is provided by the story of the Chinese envoy Kan Ying. Had he succeeded in establishing direct contact with the Roman Empire, the Parthians would inevitably have been the losers. But Kan Ying believed the

Drawing of an armored Parthian horseman, a scratched drawing from Dura Europos, the ancient city on the middle Euphrates. It was conquered by the Parthians around 100 B.C. Later they defended it against the eastward-pushing Romans and thereby probably prevented direct contact to China.

horror tales they concocted and returned to the Middle Kingdom in 97 A.D. without having accomplished his mission. The Parthians had once more proved themselves to be the worthy representatives of a monopoly. The parallels to modern times are not to be overlooked. The Persians, however, were the first to practice a method to which industrial firms today still resort, as we shall see later.

The reader may be tempted to gather that invincible difficulties for the exchange of goods must have existed at the boundary sectors of the Roman Empire and its declared foe, but this is far from true. Arrangements were made so that both sides reaped great benefits from the trade. Such contracts generally depend on the level of civilization of the contracting parties. Here two highly developed States faced each other on the Euphrates to establish a *modus vivendi* and both sides proceeded with equal efficiency. Under the benevolent patronage of Rome and Parthia local chieftans (otherwise quite unimportant) were able to enjoy their independence from both neighbors. Today we would say that they "coexisted" between the super-powers. Syrian and Parthian traders exchanged their wares in these buffer states. In the markets they forgot political differences and concentrated solely on business. Where and when buffer states of this kind could exist depended on many factors. For this reason many different routes led to Syria, Egypt, and Asia Minor from the area around Baghdad (where one could have erected a plaque with the inscription "End of the Silk Road").

Now the silk caravans drove through the northern regions of Mesopotamia and directly through Armenia in the direction of Constantinople. The silk-weaving mills in Syria and Egypt—we recall that the raw silk from China was processed in the Levant—increasingly received their raw material by way of maritime trade. But here let us dwell briefly on a factor that has barely been touched on up to now: the price of silk.

There were two especially dangerous nodal points along the Silk Road: there, where Turkoman or Hunnish spheres of interests clashed with those of China, hence in the Tarim basin, and in the Parthian-Syrian border region. In the West highly civilized peoples, who conducted their trade through constantly changing buffer states, confronted each other. Farther to the East, the levels of development of the contracting parties were too different for arrangements of this kind to be of long

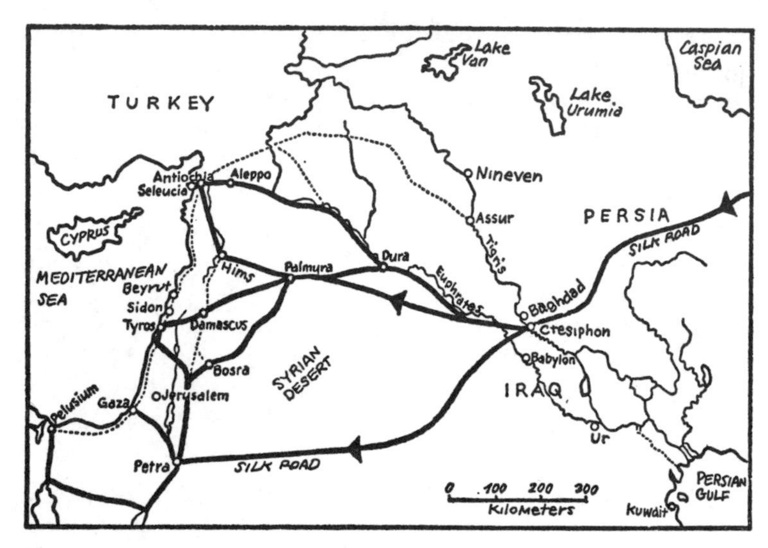

The Silk Road forked at Ctesiphon (south of Baghdad) and farther west again at Palmyra.

duration. The Chinese armies only seldom achieved full control over the Tarim basin; more often than not Mongolian lords decided whether a silk caravan would be allowed to pass through and how high the charges would be, or whether the whole cargo—which after all came from a hostile country—should not be considered as legitimate booty. In addition, the topographical configuration of the Tarim basin fostered the development of individual, independent regions. These small states or more precisely, tribal territories were separated from their neighbors by impassable regions. From the narrow perspective of their own pasture land, the petty chieftains only seldom glimpsed the advantage to them of a safe trade route. So the caravans were repeatedly compelled to buy a new and costly armed escort.

Truly the risks in silk trade were extraordinarily high. No

wonder then that the bales which had started out on their journey from Ch'angan commanded a respectable price at the other end of the Silk Road. Since an emperor was ruling in Rome the consumption of silk there had risen to astronomical heights. Items of clothing made of silk now also formed part of the new consumer craze. It certainly was expensive and the ladies of Rome–for that matter the men too–could never get enough of it. But what did they have men who conquered the world for? Those who with sword, trade carts and legal code ruled over countless peoples, whose ships, unchallenged, sailed into all parts of the Mediterranean? And they paid whatever the weaving-mills of the Levant demanded. When we read that in Rome silk was weighed with gold that is, no doubt, to be understood only symbolically. The really first class gossamer-thin fabrics were even much more expensive. Not for nothing did Pliny the Sigher ejaculate: "How much we must pay for the beauty and luxury of our women!"

But with what was Rome actually to buy the endless stream of silk? The value of the wares that left the Roman Empire never equalled that of the imports. Around the time of Christ's birth the yearly deficit ran into the millions. The difference was made up through precious metals, primarily gold. Thus a stream of gold bucked the flood of silk. The expedient did not work for long. Soon coinage metals became scarce in Rome, while it accumulated in the East. We know from the time of the interregnum of Wang Mang (9-23 A.D.) that he hoarded 150 tons of gold. This was an unimaginable amount for those times, and added to this were the silver supplies, which were greater than all the coinage metals circulating in the West. We have every ground, however, to assume that this gold did not come from Rome. Practically no Roman gold coins have been found in China, although this would be expected if the gold that paid for silk had arrived in China. But this was not at all the case; the gold remained with the middlemen who, to a great extent, exchanged their agricultural products for silk. It was there among the middlemen that we must look for Roman gold pieces and it is there that we also find them. The most comprehensive treasury of coins came to light in modern times in India, where huge earnings were made in the maritime trade with Rome.

Rome could not afford this outflow of gold over the long run. Tiberius forbade the Romans to wear clothes fashioned of silk. It had no effect because silk had long become a vital necessity. The attempt to increase the flow of goods from the West to the East also failed. At the eastern end of the Silk Road, where the population met almost all its needs from the native economy, there was no great demand for the products of the Mediterranean world.

From the time of Tiberius, for several centuries thereafter, Rome found herself in the same situation as England in the middle of the nineteenth century. Imperial England forced the export of opium to China through armed force and thus balanced the trade deficit caused by her own tea imports. The Roman world empire could not seize this opportunity eighteen centuries before the opium war, since the necessary technical knowledge was not sufficiently developed.

Another way out presented itself quite simply. Who was primarily responsible for the exorbitant silk prices? Who pocketed enormous profits and never at all thought of lowering the price when less expensive wares were brought in? The problem was to get around the middlemen–which was easier said than done. It was altogether possible to dodge a middleman on the land route, but then the trader inevitably stumbled onto another overlord's territory and was forced to request the service of another middleman. The silk trade could dodge only those with an inordinate greed for profits, and that not always. As already described, a route parallel to the Silk Road ran north of Bactria. The caravans drove along there when the costs of the direct route through the Pamirs rose too high. Another secondary stretch led through the desert south of the Aral Sea and directly across the Caspian Sea to the foot of the Caucasus in the direction of Trapezunt. These silk transports no longer touched upon the Parthian kingdom but, they were fleeced by the no less profit-greedy Armenians.

From the Tarim basin another route stretched directly across the passes of the Hindu Kush into the Indus valley, from there downstream and along the sea route either into Mesopotamia or directly into Egypt. The one or the other route was preferred according to the local conditions and hence according to the projected costs.

Whatever happened, the flow of silk from China into the Mediterranean sphere only seldom broke off, but often it had to transfer its channel. We can compare the silk trade to a self-regulating system that functions according to the principle of least force. Could not the traders have kept the middlemen completely out of the picture and fetched the coveted raw material directly from the source? That was not feasible by land. The world had already been divided up and all that the traders would have accomplished thereby was to run into other middlemen. Yet the sea was there, no one could claim its possession or demand tolls for passage. How did things stand then with the sea route to the country of the silk makers? In terms of ship building all the prerequisites for long sea voyages existed. The Roman galleys were used in wars, as is generally well known. Less often do we see pictures of the sturdy trade sailing vessels that followed in the wake of the galleys and that could accommodate considerable amounts of wares in their round bellies. For Rome and Greece the Red Sea constituted the natural point of departure for maritime trade in silk. The goods to be exported arrived in Alexandria. From there they were placed on the land route to the port of Berenice where they disappeared under the decks of the merchantmen and set out on their journey along the coast. The course led through Bab el Mandeb in the Gulf of Aden. There the wares were unloaded.

For many hundreds of years the captains always remained in sight of the Arab peninsula since they feared the open sea that offered too few aids for taking bearings. In this way, always along the coast, the journey to India took a rather long time. In the last centuries before the Christian era the trade reached as far as the mouth of the Indus and was very lively even beyond this point. Later the captains developed more confidence in themselves and in their ships and kept the course increasingly farther from the land. Finally, probably about the time when Augustus ascended the Roman throne, the monsoon blew the first merchants across the open Arabic Sea up to the Indian coasts. Thus the voyage was considerably shortened. Formerly it had taken two months to arrive along the coasts of the mouth of the Indus, now that crossing from the port of Aromata–the name clearly describes the cargo with which the ships to a great extent were laden–at Cape Guardafui to Bombay was negotiated in less than 30 days.

For us it is simply something to consider, but for the mariners it actually represented an extraordinary accomplishment, an enormous conquest of their own fears and anxieties. At that time the only navigational aids were the stars in the sky; it was practically impossible to establish geographical longitudes. So long as the ships remained in sight of the coasts or at most took a short cut across an inlet, the captains determined their positions with the help of landmarks, and on the open sea they had to trust themselves to the monsoon. Should this even once unexpectedly change direction, they were hopelessly lost on the shoreless sea. On the other hand the open sea offered greater security than coastal navigation. Pirates were not a problem on the open sea and the ships and their cargoes were safe from shallows and sea breezes. In the last analysis, however, all these arguments were not overiding; the profits to be gained were the all-determining consideration with respect to the sea voyages to India. The traders and ship owners earned so much that they practically took any risk, known or unknown, in order to get their hands on the coveted silk quickly and without cost-hiking middlemen. Luckily they could also combine the traffic in silk with the spice trade.

India herself did not produce the silk, but received it over a second Silk Road which linked the upper course of the Yangtze Kiang with the Brahmaputra. The silk was brought along the Ganges plain into the Indus valley and from there downstream to the waiting ships. Between the second and third century the

Indians almost exclusively handled the maritime trade between China and the Mediterranean world. From then on, until the ninth century, Chinese junks took over a great part of the transport. These attractive ships offered all comforts, even for the travelers on board. They had a water displacement of up to 3000 tons and crews of from 300 to 400 men. Watertight sails made them unsinkable. Travelers enjoyed cabins with their own bath—a luxury that was not to be seen again on ships for more than a thousand years.

But let us go back to the Roman Empire, whose balance of payment was never equalized because of the costly silk imports. The high prices on the land route stimulated the maritime trade, but this hardly lowered the price of silk arriving in the West. Nevertheless the Persians, who had replaced the Parthians in the third century, could not let prices soar sky-high, since the regularly arriving Indian merchantmen acted as a restraining factor. In the course of time the Persian silk dealers perceived the danger threatening their lucrative silk business if considerable amounts of goods were to be shipped directly from India into the Mediterranean region. They possessed no fleet that could cut off this trade or at least exact tolls from it. What to do? They watched as the competition took an ever increasing share of the trade for itself.

They sent agents to collect information on the situation, and in particular to find out along which routes the ships sailed. Before long the Persians were fully informed on all details. So all the ships sailed through the Gulf of Aden, through the Bab el Mandeb? Did not the Abyssinian kingdom of Aksum rule on the western shore? And were they not indeed the same gentry who played an important role in the spice trade?

We can imagine how the immensely rich Persian silk traders deliberated problems, and viewed questions from all sides: the solution was obvious. Does not one hand wash the other? If you leave the silk trade alone, I won't dabble in spices. An international cartel took shape. Persian merchants traveled to Aksum and presented their plans to the local merchant elite.

The Aksumite kingdom (from which the Ethiopian later

emerged) had acquired constantly growing influence in the outlet of the Red Sea ever since the first century. In the meanswhile the Aksumites had taken over (or rather inherited) the role of their relatives, the Sabaens who had gradually sunk into insignificance. In the beginning Indian ships took on supplies exclusively in the port of Adulis; the city of Aksum lay further inland and probably they also changed their wares there on occasion. Aksum acquired power to the same degree as the Roman Empire lost it. In the fourth and fifth century no ship could pass the Bab el Mandeb without paying a toll beforehand to Aksum. Earnings came primarily from the spice trade but Aksum was also well established as an middleman state for the silk trade. In the sixth century when the Persian merchants set forth their proposals for a mutually beneficial cartel, they found willing ears. The profits from the spice trade showed signs of declining since it was being conducted more and more along the land route, through Persia. The merchant-gentry quickly reached an agreement to divide the spoils: Persia would raise the customs duties on spice so high that the land route was bound to stagnate in a very short time. Aksum promised to do its share by preventing silk from reaching Byzantium on the route through the Red Sea. (Byzantium had become the principal consumer in the Mediterannean world.) The partners had each other in hand and could make money only by sticking to the terms of the contract, which they did. Persia drove silk prices higher, while Aksum did the same for spices. Byzantium was fleeced. The silk trade had been a state monopoly in the whole of the Eastern Roman Empire since the fifth century. The demand for costly clothes rose sharply with the unfolding pomp of the emperor and of the ecclesiastical princes. Byzantium also functioned as middleman state for the rest of the Mediterranean world and it had the greatest interest in importing silk at the lowest possible price. But this was precisely what the cartel prevented.

In 540 A.D. Emperor Justinian fixed the highest price at which the government buyers in Persia were supposed to buy.

The Persian merchants merely laughed: "Either you pay our prices or you'll have to look around elsewhere!"

Justinian's traders traveled to Aksum. There their counterparts shook their heads ruefully: "For some time the maritime trade in silk has been interrupted; even we get our wares from Persia. But we could be of service to you with spices."

Byzantium without silk? Inconceivable. Gnashing their teeth, the traders bowed to the conditions imposed by the Persians. But this time the middlemen had pushed their advantage too far, the otherwise so willing cow no longer wished to be milked and be kicked to boot. Justinian decided to build up sericulture in the Eastern Roman Empire and thereby become independent of imports. The situation was favorable. Christian missionaries were traveling everwhere in Asia; they founded churches and monasteries, and were quite knowledgeable about the individual countries. There were also Nestorian church communities in China with their own bishops who maintained contact with Byzantium.

No Chinese law is known to us prohibiting the transfer of the knowledge of sericulture abroad. Even less is there any record of a decree prescribing the death for the export of the eggs of the silk worm. These are fairy tales, no doubt circulated by those who wanted to bring the eggs abroad. The greater the danger, the higher the remuneration. For geographical reasons China hardly came under consideration in the matter of exporting eggs. The distance was all too great and there was no certainty that the precious cargo would survive. But it was altogether unnecessary to procure the eggs of the silkworm from the Middle Kingdom; the inhabitants of Indochina also reared silk, albeit not to such a great measure as their eastern neighbor.

A number of versions exist purporting to explain how sericulture was brought to Europe. The Byzantine historian Procopius, who really should know, reports that "Indian Monks" who had lived for a long time in the Far East had proposed taking over the task of importing silkworm eggs to the emperor. After the emperor had promised them a rich reward, without which they would not have taken the risk, they set out on their journey. After several years they came back with the coveted eggs. Scholars today incline to the view that these Indian monks were Nestorian Christians from Persia who procured the eggs in Cambodia. This view has not been contested, but the fact is that silk breeding began in Syria in 553 A.D.

We should not lend credence to the fairy tale that it sufficed to bring the eggs to Byzantium in a hollow pilgrim's staff. It just wasn't that simple. After all silkworm rearing was a complicated affair which promised success only in the hands of real experts. Accordingly not only the eggs of the silkworm arrived in the Byzantine Empire in the middle of the sixth century but the necessary technique, the "know-how" was also imported.

Thirty years later Syria produced so much silk that practically no more had to be imported by way of Persia. The Persians got their comeuppance for their immoderate greed for profits. Now prices fell so low that the profit dropped from a percentage of many hundreds to ten per cent, but it did no good. The long transport routes and the considerably heightened risk–Mongolian peoples once more controlled a long section of the Silk Road–increased the price of Chinese silk. Traffic on the Silk Road shrank–without completely disappearing, however.

Byzantine sericulture meanwhile had developed spectacularly and Byzantium itself now exported silken fabrics. After Byzantium lost its Syrian possessions, silk-rearers shifted their operation to Greece. From there the knowledge of silk-making radiated to Italy by way of Sicily, and further to Spain. Silkworms now hatched from their eggs all over Europe, spinning their cocoons and delivering the coveted raw material for the silk-weaving mills. Chinese silk could not offer competition as long as the control over of the Silk Road kept changing hands and the controlling lords kept exacting high tariffs from the caravans.

The flow of silk directly across the Eurasian continent first swelled once more to its old vigor at the time of "Pax Mongolica" when the Great Khan ruled the greatest empire of all times that stretched from the Yellow Sea to Budapest (thirteenth and fourteenth centuries). During this time silk caravans plodded along unmolested from China to the West. There were no robbers and bandits; the Mongolian tomans had pitilessly pursued them to their lairs and wiped them out. The Silk Road was so safe, day or night, that a lost bale from a caravan could be picked up intact on the return journey. Moreover, China now no longer primarily simply exported raw silk which was then processed in the Levant: Chinese patterns were now in demand in Europe. Thus the rolls of silk arrived directly from Chinese weaving mills to the tables of the merchants and tailors of Europe.

But not only wares and goods now passed over the Silk Road, once again pulsating with life. For the first time European travelers made their way along the full length and saw the Middle Kingdom with their own eyes. In 1260 Nicolo and Maffeo Polo arrived at the court of the Great Khan Kubilai. After a lengthy sojourn they traveled back to their Venetian homeland and soon set out again for Chambalic, this time accompanied by Marco Polo. All three men occupied high State offices and Marco Polo in particular, as a personal emissary of the Kubilai, traveled extensively over the country. From his pen came the first thorough description of the Silk Road, including the landscapes and cities lying to the right and left of it. After a time so many envoys, merchants, and missionaries streamed to the Far East that in 1340 the Florentine Francesco Balducci Pergolotti wrote a kind of travel guide. Besides such miscellaneous useful information as where one would do best to let one's beard grow, where money could be exchanged at the most advantageous rates, and how a merchant could obtain a pass in order to be allowed to use the dispatch rider service he also reported that the Silk Road was absolutely safe at any time.

With the decline of the Yüan (Mongol) dynasty in 1368, the life of the Silk Road also faded away. The Yüan had been open to all alien influences and ideas, but now the Ming practiced a policy of the closed Chinese Wall. The Middle Kingdom encapsulated itself from the outside world; only the Jesuits succeeded in setting foot there for a time, after 1601, when Matteo Ricci was allowed to journey to Peking.

Today the Silk Road is again cut up by numerous borders. Gone are the times when travelers from Syria could approach the Middle Kingdom on slow but sure donkeys, camels, Bactrian dromedaries, horses or ox-carts. The autos, buses and trucks of the twentieth century are indeed faster, but they are not allowed to cross certain borders. The distance between Europe and China has grown again, the old bodies of knowledge and standards are no longer valid, and only seldom does information trickle through the closed Chinese Wall. It was easier to learn the truth about what was happening in China when Ch'uan Sun observed the Roman mercenaries beyond the Talas than it is today in the age of space travel.

13.

Columbus was Chinese

ONLY twenty years from now the New World will be making preparations to celebrate the quincentenary of the discovery of America. The name of Christopher Columbus will again be on all lips. Replicas of the "Santa Maria," the "Pinta," and the "Nina" will follow the original course across the Atlantic as recorded in the log books. If they sail with a favorable wind, they will see on the horizon the same island that brought the lookout on Columbus's flagship several gold pieces. Which island? This is only the first of a long chain of questions that surround the discovery of America. When we consider that the Genoese in the service of Spain cast off at a time when numerous chroniclers were recording anything that seemed noteworthy, the many unanswered questions should be a source of astonishment to us. After all it was not the voyage of some daredevil in the grip of a fixed idea, commanding three slave-trading ships with desperadoes for crewmen. On the contrary, the expedition was planned and equipped as thoroughly as it was possible to do at that time. The financial situation of the Spanish throne was not so rosy that the Chancellor of the Exchequer would have airily invested several hundreds of thousands of ducats in a project that did not offer at least a chance of success. The ruling house of Castile had no interest in a voyage of discovery for the purpose of broadening mankind's geographical horizon. The only thing that could open the lid of the royal cash-box was the credible-sounding promise that each gold coin expended would be replaced by at least two counterparts, and within a relatively short time.

The Chancellor kept an account book of the costs of the voyage, the equipment was thoroughly discussed in countless letters and petitions, the aims and missions established in contracts and, finally, the result was most accurately investigated in court proceedings and preserved for posterity. The voyage of Columbus to India might then seem to present a model example of written history, a voyage that we can reconstruct in all its details. Yet this is not so by a long shot. There is probably no event in world history about which so much has been written, speculated, disputed, and fantasied so thoroughly as this expedition westwards across the Atlantic. Certainly, Columbus inaugurated regular contact between Europe and America, a contact which has since constantly grown stronger. After Columbus world history entered upon a different course, and it is not for nothing that school history texts mark 1492 as the end of the Middle Ages. Thenceforth begin the pages of modern times. There are fewer analyses dealing with the extent to which Columbus's discovery influenced the further course of the world than there are speculations on the discovery itself (which fill whole libraries).

We cannot even clearly answer the first question posed at the very outset: which island provoked the cry "Land ahead!" from the lookout in the crow's nest? It was probably San Salvador, but this not certain. Yet this is a simple question compared to others that surround the Genoese. How, for example, did he manage to persuade Their Majesties Ferdinand and Isabella that he was the only man to whom they could entrust an expedition of this kind (and a very costly one to boot)? He was not a Spaniard by birth, and this was a disadvantage in tradition-bound Castile. He had not distinguished himself through previous maritime expeditions, and his reputation as mariner and navigator did not tower above his contemporaries, many of whom had sailed around the African continent in the direction of the Cape of Good Hope. Natives of neighboring countries were always found on the Portuguese ships that Henry the Navigator and his successors sent out. Would they not have been natural candidates for a command of this kind? Nor did Columbus have any new ideas to offer. The cosmologists at the royal court had already long asserted that one could find the sea route to India by pushing forward across the Atlantic in a westerly direction. The captains and adventurers Fernam (1457), João Vogado (1462), Fernam Dominguez de Arco (1484) and Fernam d'Ulmo (1486) had already received from the Spanish throne covenants concerning islands and countries to be found in the Atlantic. D'Ulmo's charter even expressly mentions that the expedition would look for *tierra firme*. This is a clear indication that, in the last analysis, India was bound to be found if one only sailed far enough.

The cosmographers (above all the Florentine astronomer Toscanelli) were not the only ones to be convinced that land was bound to be found on the other side of the Atlantic. Reports, directly or indirectly supporting these contentions, piled up in the archives. Already several generations had told of Irishmen who lived in a distant land beyond the Atlantic. And what about the legends of the northern countries? Perhaps they contained a kernel of truth and the Vikings on their voyages westward had actually seen—and even set foot on—land. What about the reports of antiquity? A happy people was supposed to live beyond the pillars of Hercules. Contemporary mariners also reported signs that indicated the proximity of land when storms on the Atlantic had driven them towards the West.

No, this idea of Columbus was not a novelty at all. What secret did Columbus possess then that fetched him the command of an expedition at State expense and guaranteed him extraordinary rights and privileges in the event that he discovered land? (That the king in the end reneged on his promises is another matter.) He was to become the viceroy of the new possessions, and was even to be allowed to keep ten per cent of all the income for himself; the crown granted the same rights to his children as well. Truly, Columbus must have revealed an important secret to the Spanish king and queen, but we do not know what it was. Perhaps Columbus was just a consummate actor who convincingly gave the impression that he possessed this secret knowledge. Did he actually want to sail to India and China? Much evidence suggests that the rich Spice Islands were his goal. Thus he drove the fleet forward pitilessly in a westerly direction, even though signs cropped up en route that land might be near. He did not want to lose time by cruising around. Yet no sooner was land sighted (though it turned out to be only an island) than he exhibited no further interest in looking for the mainland in the West. According to his calculations these were the Spice Islands. What should he do on the coast of the Asiatic continent? For him the attainable treasures were the Spice Islands. As an Italian he was familiar with Marco Polo's reports concerning the populous and mighty countries of Asia. In Spain, where the Islamic armies had just been expelled, the reports of Arabic world travelers, in particular those of Ibn Battuta, were not unknown. The Genoese must also have known about them. He was certainly not unrealistic enough to hope to be able to take possession of this empire for the Spanish throne with three tiny caravels and a handful of soldiers. Marco Polo had written about crack armies numbering in the hundreds of thousands. Let us just imagine that Columbus actually had arrived at the Asiatic coast, had called at the next port and informed the authorities that he was taking possession of the country in the name of the Spanish throne according to contract. Even if the port commander had granted him an audience, Columbus would have certainly landed in prison for unbecoming conduct and vagrancy. In short, neither India nor China nor Japan shaped Columbus's course when he set sail from Palos a half hour before sunrise on August 3, 1492.

There is another, perhaps much more interesting theme in connection with the discovery of America. What did Columbus actually find in 1492? Let us cast a glance at the double continent stretching from Alaska up to Tierra del Fuego. Without dwelling on the bloody history of how the individual peoples were subjected and exterminated, let us consider these civilizations as still undisturbed by the white man, they emerged, flourished, and declined.

At first glance these people appear quite uniform in appearance. They are all Indians. Among them are hunters and fishers, cultivators of the soil and cattle breeders, village and city dwellers. The outer picture of the inhabitants is common to all: smooth black hair, tinted complexion, prominent cheek bones, dark and frequently almond-shaped or slanted eyes. Everything indicates that a more or less uniform aboriginal population settled here, from which many groups developed (some more slowly than the others), as was also the case in Europe. Do we see here a true model of development from hunters and gatherers to highly civilized city dwellers?

On closer investigation of these cultures, we find peoples and customs that simply will not fit into this uniform image. This point is given scant attention in all the discussions of the secrets surrounding Columbus and of the history of the conquest of America. There is not a great deal of research in this area. In good part, moreover, it has been limited to the investigation of details. Buried in the storerooms of libraries, these findings await the interest of future generations. Let us follow these traces. Perhaps we can give them up after a short time and again concern ourselves with Columbus, if they fit the pattern of consistent gradual cultural development. On the other hand, if this is not the case, we may have to resort to speculation in order to understand this intrusion on the American continent. Three cultural areas merit our attention.

The Northwest Coast

The range of the Rocky Mountains rises in northwestern America to an elevation of more than 18,000 feet. Their ice-covered slopes reach down to the Pacific and form an almost impassable barrier to the rolling North American country lying farther east. A narrow strip of coast runs between the sea and the mountains which even today is accessible only by ship. Numerous islands line the coasts of the southern spur of Alaska and the Canadian province of British Columbia and break the ground-swell of the Pacific. The warm Alaskan current coming from Japan provides for an even, relatively warm climate. The Gulf Stream plays a similar role with respect to northern Europe. The westerly winds lose their moisture on the coastal highlands and the resultant warm rain fosters a luxuriant vegetation. The numerous brooks and streams abound with fish. The first trappers reported that at the time when the salmon were running a man could cross the streams without getting his feet wet. The fjords, inlets, and protected reaches of water belong to the richest fishing grounds of the world. In the pine woods of the western slope of the Rocky Mountains live wapiris, deer, stags, elks, goats, mountain sheep, pumas, brown bears, and grizzlies, to mention some animals whose very name makes a hunter's heart beat faster. In short, a paradise for human beings–albeit in a most unexpected and remote spot accessible only by boat.

The Tlingit, Haida, and Chimmesyan are foremost among the Indian tribes living there. They stand on a significantly higher stage of civilization than the neighbors with whom they share the coast land. They live by hunting and fishing; agriculture is unknown to them. Although they cannot make good pots, they are masters at working metals and famous for their weavings. A strict caste system divides them into three groups. The noble class has nothing much to do besides ruling and spends its time with artistic work and festivals, not–as we might presume–with waging war and making conquests. If they want to increase their prestige with their neighbors, they make show of their wealth in the following manner. One chieftain proves that he has more than he will ever need by throwing tons of fish oil into the sea, burning his stores of smoked salmon, throwing dried meat to the dogs, cutting into shreds the most beautiful animal pelts, and hurling the most precious copper-pieces into the sea. The opposing party answers the challenge with an even more wanton extravagance—or remains shamefacedly silent. This astonishing competition for preeminence finds its climax in the *potlach*, a festival in which the most prestigious tribesman divides all his possessions. This, of course, in great part is at the expense of the members of the middle caste who must produce all that which is so prodigally thrown away. Their lot is eased by the slaves who stand on a rung below them.

But it is not only this exceedingly unique way of acquiring status and prestige, or saving face, that distinguishes this tribe from its neighbors. In double canoes, up to 75 feet in length, they undertake extensive voyages along the coast and also venture far out into the sea. They mark the position of their large wooden houses with totem poles, a confusing vertical jumble of stylized human and animal figures. The material for these poles, the house-high Scots pine, grows right in front of the door and the upper class has plenty of time to devote to sculpture.

If they draw, weave or sculpt an animal, they show it from both sides. The artist depicts the right profile of the bird and then fits the left profile to it. Left and right profiles are united by the line of the back. This is called a bilateral representation.

These Indians have, of course, taken over much from their more primitive neighbors who, in turn, have learned much by observation. Nevertheless their culture is unique on the American continent; we find nothing resembling it either in North, Central, or South America. This "Northwest Indian Culture" was first discovered in the eighteenth century by Russians, Spaniards, Frenchmen, and Englishmen. Today is has already

A splendid example of the bilateral animal representation of the Haida, a tribe on the Northwest Coast of America. Illustrated here is the famous Thunder Bird.

become difficult to identify features and customs still uninfluenced by the white man. Here in the northwestern part of the American continent is a section incongruous with its environment, a somewhat remote and little-known region. The second alien group leads us to a more familiar area, Central America.

THE OLMECS

South of the Tropic of Cancer the bridge of land between

North and South America is drawn in tightly at one place before it widens into the Yucatan peninsula. The Isthmus of Tehuantepec separates the Atlantic Gulf of Campeche from the Pacific Gulf of Tehuantepec, but also links the two seas at this spot. The spurs of the Sierra Madre del Sur consists of hardly more than hills; here nature facilitates the nexus between the Atlantic and the Pacific. On the Atlantic side, between Veracruz and the border of Yucatan, we find wide stretches of lowlands. Extensive jungle regions stretch deep into the interior behind the coastal lagoons. It is an area into which modern man would not venture without a compelling reason; for the native peoples however, it is a tropical paradise. Everything that man requires for life grows here. Enough animals roam through the swampy primeval forests to satisfy the hunter. No one would ever think to look for the remains of an advanced culture here, yet at one time this was the site of the civilization of the "Dwellers in the Land of Rubber," of the short-lived kingdom of the Olmecs.

Since the turn of the century signs have multiplied indicating that once men must have lived on the plains of Tehuantepec who would have had little in common with the Indians now vegetating there. Artistic jade sculptures came into the hands of archeologists. Curious, they pushed deeper into the interior, where they found three cities completely overgrown by the jungle. The best known Olmec site is "La Venta," lying on an island in the Rio Tonalá. Pyramids, temples, steles, tombs, and monumental sculptures were laid bare by machetes and trowels. The form of the pyramids and steles greatly resembles that of the Mayas and the Aztecs, yet nowhere in America does this sculpture have a counterpart. The heads carved out of stone are particularly striking. From crown to chin they measure from 4.70 to 9.85 feet. The faces are realistically depicted. The features indicate a light Negroid strain and do not seem characteristically Indian at all. Artists must have gotten the stones from quarries that lie more than sixty miles away as the crow flies. The raw material could have been transported only on rafts.

In the tombs the archeologists found jade figurines wrought

with a consummate artistry. This material was also not easily accessible to the inhabitants.

The Olmecs already knew the art of calendar-making and possessed an alphabet—the oldest in Central America. The earliest finds can be ascribed to around the time of the birth of Christ; before that we find only traces of primitive basket-makers and gatherers. In other words, the Olmec culture appeared quite suddenly and at a level of perfection, without any preliminary stages. (The oldest jade pieces are most impressive of all.) Where did the Olmecs come from? We do not know. A highly developed culture suddenly sprang into being and after a few hundred years was destroyed by invaders from the outside. We find the influence of the Olmecs in all Central American cultures, especially with the Mayas and the Zapotecs.

THE CHAVIN CULTURE

For the moment let us leave the Olmecs and continue on to South America. There, where the sub-continent swells out westward we find another alien element in the cultural pattern of America. It lies in the north Peruvian Andes: the Chavin culture, named after one of the most important find-sites, Chavin de Huantar.

Some may ask why one should concentrate on this culture when after all there are many Peruvian finds of a much more spectacular character—the Incas, for example. Where the Incas came from and on whose culture they built has been roughly sketched out by now. The origins of the inhabitants of the Chavin epoch, on the other hand, is a riddle. Around the tenth or ninth century B.C., they were suddenly there. They cropped up as spectacularly as the Olmecs did one thousand years later at another site, with a fully complete civilization which thereafter barely progressed. What is particularly astounding is that these people knew no metal but gold. They were master goldsmiths, apparently from the start, since we have no evidence of any preliminary stages of their art. The broad double spiral stands out as the leitmotif of the ornamentation on jewelry and stone sculptures. It is also found on the side of a strange figure carved in stone: a beast of prey with bared teeth, probably a jaguar, carrying a cylindrical vessel on its back. In the same area investigators found another representation of the same animal in bas-relief. The double spiral is again clearly recognizable. The scaled tail of this awe-inspiring creature seems to come from the sea.

A few hundred years later the Gallinazo and Mochica cultures appeared in the same region. They too seem practically overnight to have made achievements, the evolution of which in the Old World required millennia. This is most evident in their grasp of metallurgy, for archeologists found vast copper objects of the highest degree of perfection. In the case of stone sculpture, one can dispute which intermediate phases characterize true art of a highly developed culture. This is not so with metallurgy. Hammering, annealing, melting, extracting ore by melting, casting, and alloying, for example, are the intermediate steps taken by the cultures in Mesopotomia and on the Indus. But in northern Peru, one of the most difficult casting techniques from the outset—casting by the cire-perdue or "lost-wax" process—was known.

These people learned about bronze at a much later date: in their environment there were no deposits of zinc (the additional component of bronze next to copper). Thus a scholar has rightly described the epoch of the Gallinazo culture as a Bronze Age without bronze. Since we are at the moment dealing with techniques of working metal a rare custom rates mention here: for beauty's sake, the natives carved broad grooves into the front of their teeth and filled them with tiny gold plates.

Metallurgy was not the only technique that experienced an early peak of development in the Central Andean regions; this people also perfectly mastered the art of weaving. The resist-dyeing techniques for coloring cloths, *plangi, ikat*, and *batik*, were known. These are methods in which a part of the fibers is protected by wax coatings and as a result "resists" the dye. The textiles patterned in this way are unique on the American continent. Many other skills that distinguished the people of this region from their neighbors could also be listed, such as knowledge of knots, sailing rafts, musical instruments, and hunting methods. In short, in the northern Peruvian highland, also known as the Central Andean region, a highly developed civilization appeared around the tenth century B.C. for which no preliminary stages were found in America. The subsequent cultures also are characterized not so much by gradual development as by sudden thrusts. This picture remains unchanged until shortly after the beginning of the Christian era.

This third culture, seemingly out of step with the general development on the American continent is quite congruous with the aforementioned cultures in British Columbia and Mexico. What is common to all of them is that they are very clearly distinguished from the civilizational stages of their neighbors and that no specific independent development is demonstrable.

The reader may not find this so amazing. He may point out that America is an immense continent with an enormous population which migrated across the Bering Strait several thousands of years ago. There was plenty of time for independent development, and particular forms do tend to crop up when local conditions require them. Finally, it seems quite obvious that the Indians were responsible for these early civilizations on the Northwest coast, in Mexico and northern Peru. But is this accurate? Before we concern ourselves with this problem more closely, we must make a small detour in order to learn a little more about these American aborigines.

All of us have been very familiar with America's first inhabitants, the Indians, since the days of our childhood. At first sight, we would classify the Indian race as related to the Mongolian. In fact, all evidence points in this direction. The continent was still uninhabited at a time when the peoples of Asia were already treading on each other's toes and fighting each other for possession of the best hunting grounds. Fortunately enough, just then the ice in the north receded and set free a land bridge between Asia and America. The Mongoloid ancestors streamed into the still uninhabited prairies and highlands, settled down there and developed their own, independent cultures, as the land was soon cut off by the Bering Strait. Today the more or less pure descendants of these migrants may be seen in the reservations of North America. The reservations of other Indian tribes were not determined by administrative borders, but by nature and the will of later conquerors. (Examples are the barren regions of Tierra del Fuego, the high Andes, or the Amazon basin.)

A closer look at widely dispersed Indians will reveal, surprisingly, that only few accord with our classic image of the Indian. There are tribes that exhibit pronounced Australoid traits, in others a Negroid strain is unmistakable (from which we generally conclude that these particular tribes must have mixed with the Negroes imported as slaves). Still other tribes are hardly distinguishable from deeply tanned Caucasians. There is no homogeneous Indian race, and we must drop the idea that a single group of Mongoloid people crossed over the Bering Strait and thereafter broke off every contact with the Old World.

Again, the reader may deem it possible that these special traits developed in America independently of the rest of the world since a rather long time span was available for such progression. In 1961 Stalker found on the Mississippi the remains of a human being to which, in 1969, he ascribed an age of at least 37,000 years. He did not exclude the possibility that they might even be older, going back perhaps as far as 60,000 years. These data are quite in keeping with other finds in Idaho, Texas, California, and Nevada. This time span—so brief in comparison with man's evolutionary history—is by no means long enough for the independent development of different racial traits. Apart from that it would indeed be amazing if in the New World the only traits that developed were those that also exist in the Old World. A revised theory of America's beginnings is in order.

The first inhabitants must have penetrated the American continent in more or less separate waves. This would provide an explanation of the different racial features. The complete scientific picture of this population movement has not yet been drawn up, but Schlesier has demonstrated eight successive prehistorical waves of immigration.

During the Ice Age a land bridge existed between Asia and Alaska. The ice held so much water that the sea level lay more than 75 feet below the present level and the continental shelf appeared on the surface in places. In the first half of the Würmian glaciation (around 40,000 to 60,000 years ago) people streamed out of the northeastern regions of China along the Pacific coast across the Bering Strait to Alaska. At that time an ice-free corridor, opening before them along the Yukon, led into the interior of the continent. Originally they must have followed the wild game that was particularly thick on the ice-free land bridge. Wooly rhinoceroses, mammoths, horses, camels, primitive buffaloes, caribou, mastadons, and musk-oxen were the prey stalked by these Stone Age hunters.

Up to now the Caucasian peoples from inner Siberia have always been considered as the first arrivals in America. This view, however, is no longer tenable in the light of new findings, since Siberia was first settled in the second half of the Würmian glaciation. It was not until the second wave that the Caucasoid groups from the Altai region found their way along the northern edge of Asia to America. However, the term "northeastern China," is used here only in a geographical sense. The people who came from there were not all Mongoloid; in fact the latest investigations show that this area was inhabited exclusively by Neanderthaloids–and a Neanderthal man is an altogether respectable candidate for the role of discoverer of America.

The second migratory wave, which began in Siberia and included Caucasians from the regions between Altai and the Aral Sea, took on Mongoloid elements when people from the

Ordos region in Inner Mongolia joined the movement. These Stone Age hunters remained mainly in Alaska and on the Pacific coast, since a new ice barrier barred them from direct access to the interior of the American continent.

It will be noted that the chart of migrations on page 278 only indicates the movement from Asia to America. Yet the land bridge was by no means a one-way thoroughfare. Between periods III and IV, a movement set out from Alaska and pushed into inner Asia. Its traces can be followed up to the region of Lena in Siberia. These migrants must have been very successful hunters, since many zoologists ascribe to them the extermination of several huge animals such as the mammoth.

It is difficult to say where the descendants of these groups are to be found today. Except for the first wave, which was able to push forward through the ice-free corridor into the interior, the members of each new migrating wave found competitors before them. The later arrivals could mix with the original population; they could drive them out of their hunting grounds and thereby push further south; or they could bypass them. These migratory movements of the peoples of America, which continued up to the last centuries before the Christian era, have been researched only in part.

Thus the first human beings set foot on the American continent some 40,000 to 60,000 years ago. As hunters they brought a very primitive Stone Age culture with them and could penetrate deep into the interior at any time. The ice barrier closing behind them was first re-opened around 10,000 B.C. to show new peoples a route to the south from Alaska. The Algonquin groups were the first to bring the art of pottery making with them from Asia in 3000 B.C. The pressure of new migrants from the north in the search for better living space led to a series of constant thrusts southwards. Thus apart from the Neanderthaloid group that was the first to arrive, settlement of Central and South America can have begun only with the advance of the Algonquins (around 3000 to 1000 B.C.). The level of culture brought along was still very low; only the Algonquins could have introduced a primitive technique of working copper. The many different waves of migrant peoples explain the variety of racial traits we find in America. These people did not bring highly developed cultures with them, yet such cultures do arise during a period in which we would expect only hunters, gatherers, and simple cultivators. Let us therefore return to our original question: can the cultures in the three regions–the northwest coast of North America, the Isthmus of Tehuantepec and northern Peru–be understood as special outgrowths of the aboriginal population of America?

The outer appearance of these people, so far as we are able to reconstruct it, tells us nothing; too many races have been mixed on the American continent. Nor is the time factor a clear-cut case. We do not know how long the Tlingit, Haida, and Chimmesyan have lived on the coasts of British Columbia. Some estimate a span of 4000 years.

The earliest traces of pre-Inca cultures appear in the tenth century B.C. Aside from the Olmecs (who built their first cities at the beginning of the Christian era) we come upon difficulties, for the first people exhibiting a stage of development beyond that of the primitive Stone Age hunters, could not have arrived in Central America before 2000 B.C. This leaves little time for a development up to the level of the Central Andean civilization. –Then there were special circumstances that accelerated this development. –But why did they not progress beyond the high point which they achieved at such an early stage? Why do we find in America mainly forms which we also know from the Old World? –That is only partially true; there are significant deviations. –Did the American Indian, then, develop all these cultures wholly by himself? –Not at all. These developments can only be understood if we assume the existence of outside influences.

This is the kind of quarrel that goes on between the "isolationists" and the "diffusionists." The former postulate a single migratory explosion from Asia after which America is to have developed in isolation from the rest of the world. The diffusionists, on the other hand, argue that this time span was by no means sufficient for the widely dispersed independent

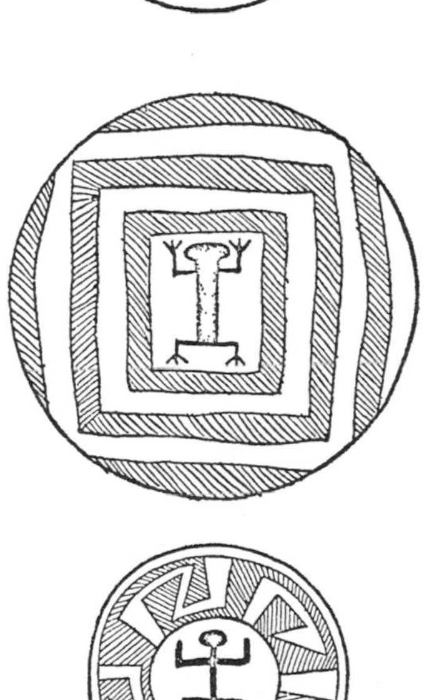

The last migratory wave across the Bering Strait probably brought simple techniques for working copper to America. Otherwise how can the similarity be explained between the black and white painted bowls from Kansu, China (top) and those from New Mexico (Pueblo Culture, center) and Arizona (below)?

history seen on the American continent. Advanced American civilizations assertedly came into being as a result of the penetration (diffusion) of other cultural currents from the outside, and the only problem is to explain where these impulses came from.

The fronts of the two camps have become extraordinarily

rigid. On the one side stands the Establishment (above all American ethnologists, archeologists and sociologists), on the other the revolutionaries under the leadership of the late Professor Robert von Heine-Gelder and his American colleague Gordon F. Ekholm. Life is not easy for exponents of a purely local, inner-American development, for certain phenomena can hardly be explained by the isolationists on a rational basis. How is it, for example, that the development of some Central and South American civilizations can best be described mathematically in terms of a discontinuous function? Here, a long period at a very primitive cultural level is followed directly by a significantly higher stage of civilization, which appears within a short period of time and hardly ever undergoes further change thereafter.

The real difficulties arise when we tackle more specific problems. How is it, for example, with the "lost-wax" process of casting metal? We cannot explain the appearance of bells cast by this complicated process in the northern Peruvian highland (and even farther north in Ecuador) in 4 B.C. by ascribing it to sheer chance. We find no intermediate links that suggest a gradual improvement of the metal casting technique. At best, only a few hundred years were available for an independent development, if we posit that the simple preliminary stages have escaped the archeologist's trowel up to now. In Mesopotamia and in the Indus valley the same development required 1500 years. The Algonquin Indians living in the region of the Great Lakes of North America work copper scraps which they can gather up in pure form. Although they have lived in this region for more than two thousand years, their technique of working copper has not yet gone beyond the second stage; they still form the tips of spears by hammering the cold metal and tempering it by repeated heating. There is no trace among them of a casting technique of any kind, be it ever so primitive.

Some arguments can also be raised against the idea of an independent invention of pottery-making in America. As Gladwin pointed out in 1947, the skills required for this art are by no means as few as archeologists and ethnologists presume. A particular kind of mud or clay must first be found in the environment. Clay can be made malleable by water, of course, but sand also must be added so that later the pot does not crack during the firing. And sand must be used in precisely the correct amount. If the formed pot—it is curious by the way that the same forms originate practically everywhere—contains too much water, it will fall apart upon drying. If it is too dry from the outset, fissures form. Finally a certain temperature range is required for the baking. If we then consider the different ways of applying color to pottery and the precautionary measures that must be observed in the process, it will be obvious that only a development stretching over millennia can lead to these processes. But in Central and South America, considerably briefer periods of time were at the disposal of the inhabitants.

Consider the intricately wrought jade figurines of the Olmecs who simply seemed to appear out of the blue. It is impossible that the Olmecs came from the simple basket-making culture living there before them. And what about the calendar system of the Olmecs, or their writing? The simple hunting and cultivating peoples who according to the isolationists are responsible for it, lacked the leisure to occupy themselves with things of this kind. Such activities would have been irrelevant to them.

These examples should suffice to underline what one of the foremost experts on pre-Columban cultures, Professor Walter Krickeberg, has written: "The oldest advanced American civilizations appear on the scene seemingly without roots, without preliminary stages: the Olmec culture in Mesoamerica, the Chavin culture in the Andean countries. This remarkable phenomenon can be satisfactorily explained only if we assume one or more impulses that worked on America from the outside. Otherwise it is difficult to understand how primitive circumstances could have endured for 15,000 to 20,000 years with few changes, and then within a span of 2000 to 3000 years storm through the whole series of stages up to advanced civilization. In the case of the two oldest advanced American cultures, there is not even a trace of this ascent: suddenly they are simply there."*

Again the exponents of a purely inner-American development will object, reminding us that these are mere fanciful speculations and asking us for proof. But is it reasonable to expect direct proof? No one today would dispute that the Vikings landed on the American Atlantic coast around the year 1000 A.D., yet what proof exists for this? There are the two Vinland sages, *Graenlendinga* and the *Saga of Eirik*, which were handed down from Iceland. Not until 1965 was a map found from the year 1440 that could be traced to the data of the discoverers. This medieval map unmistakably shows regions on the North Atlantic coast but even today the individual bays and peninsulas have not yet been clearly identified. A Danish archeologist who has been digging on the northern tip of Newfoundland since 1962 has found what appears to be the remains of a Viking colony: a whetstone, a distaff flywheel for spinning, and a bronze needle. These are meager traces for an event preserved in songs that took place a mere thousand years ago. In addition archeologists have expended much time and effort over more than a century searching for traces of the Vikings on the American continent. What, then, can we expect in the way of direct proof concerning events that transpired two thousand years ago (the emergence of the Olmecs in Mexico) or even three thousand years ago (the Chavin culture)? We cannot expect to come upon a "confession"; a chain of indices will constitute all the evidence that we can hope for to lead us to the author or authors of these events.

What external impulses produced the civilizations on the coast of British Columbia, the Isthmus of Tehuantepec, and in the Andean highlands, civilizations so alien to those of their neighbors? To answer this question, we must collect as many unmistakable particulars as possible and then look for their counterparts elsewhere in the world. To describe all the characteristics of these cultures that have been discovered would bring us far beyond the scope of this book (the reader is

**Altmexikanische Kulturen,* Berlin, 1966.

referred to the bibliography); but we shall focus on some points that are especially striking.

In the civilization of the Northwest Coast, the most outstanding feature is the totem pole driven into the ground beside the large wooden houses. It plays the same role for the Haida, the Tlingit and the Chimmesyan as did the coat-of-arms in medieval Europe: it identifies a clan or family. Each of the many clans represented itself through an animal. Bears, lynxes, and ravens are the most frequently used heraldic animals. The totem poles are covered from top to bottom with illustrations of animals and humans, whose sequence has a particular meaning. The individual squatting figures–stacked one on another and intertwined–are often abstracted to such a degree that they are represented only by an ear, teeth, or an eye.

The totem poles that still stand today all stem from the nineteenth and twentieth centuries, since in the humid coastal climate tree trunks exposed to the elements rot in less than one hundred years. Captain Cook, who sailed along this strip of coast in the eighteenth century, described these striking structures, and there is no reason to doubt the great antiquity of this art. The famous English discoverer saw something else among these coast-dwellers as well that seemed to him worthy of being recorded: iron utensils that were definitely not from Europe. Unfortunately, they too have since succumbed to the ravages of time.

As late as the end of the nineteenth century, huge double canoes (to 75 feet long) still existed that could accommodate 50 men and their equipment. Most were driven by paddle, but a sailing device was also on hand. The sails themselves were made out of thin wooden boards which could be folded like a modern Venetian blind.

We have already mentioned the bilateral animal figures among the artistic motifs. These also included two-headed dragons and representations of human faces or eyes in the joints of figures.

We do not know when this civilization came into being. Most researchers ascribe more than three thousand years to it.

No one has yet analyzed the myths and tales of this people—a definitely worthwhile if laborious project. In the main these stories refer to individual clans and usually do not touch on broader associations. There is an indication that these old accounts may be more reliable than we tend to assume at first. To this day the Tlingit describe in great detail the welcome which in 1786 they accorded to the Frenchman La Pérouse in the Lituya inlet. In short, then, the major features are: totem-poles, copper—and possibly iron—processing, sea-worthy double canoes, two-headed dragons, bilateral depiction of animals, a caste system.

The name Olmec refers more to a cultural area than to a group of peoples. Thus it can apply to several human groups, differing greatly from each other, whose unifying element is that they all lived in the same region, on the Gulf of Mexico between the Alvarado and the Los Terminos lagoons and belonged to the same culture, very different from their environment. Perhaps for this reason, we cannot reconstruct even a single trace of an Olmec old lingusitic family. Two different physical types are clearly discernible in the art that has been preserved. One type is of low stature, indeed almost chunky, with massive shoulders, round heads, broad noses, and thick-lipped mouths with downturned corners. They strike us as unusually lifelike. These colossal heads joined directly to a platform were found at La Venta along with other material. Why the Olmec artists did not sculpt the bodies as well is a mystery. The reason may lie in purely technical difficulties since there are no stone quarries in this jungle territory, and the raw material had to be hauled from afar. However, there are complete figurines in jade, among which some bear the same heads in miniature. The other physical type, as depicted by the steles, reliefs, sculptures, and ornaments is tall, with a narrow face and often slit-eyed. Some of them seem to wear a pointed beard. In Tres Zapotes a stele was found that obviously depicts a homage scene. Both of these physical types, so different from each other, jointly accept the act of submission of a kneeling figure.

The Olmecs exhibit a mastery of figural representation that is still unmatched in all of America. The figure of a sitting naked athlete (perhaps that of a ball player), stemming from the province of Tabasco, can hardly be surpassed in vitality and lifelikeness.

The Olmec cities of La Venta, Tres Zapotes and San Andrés Tuxtla, which were discovered in the jungle, are not very impressive when compared to those of the Mayas and the Aztecs. It is, of course, very difficult to erect large structures in swampy alluvial lands. All large stones must be transported by boat or–if they weigh more than 50 tons (as do some in La Venta)–on rafts from distant stone quarries. Once we take these natural givens into consideration it will not surprise us that the quadratic pyramid at La Venta is built of clay rather than stone; this is still a considerable achievement if we recall that the length of the sides is over 150 feet. The pyramid is flattened out on top and perhaps bore a shrine at one time. A stone staircase also led up to it. Unlike Egyptian pyramids, this structure did not serve as a tomb. Archeologists discovered graves under mounds of earth several yards high which covered sepulchral chambers erected out of stone slabs. Some of the dead found their last resting place in monolithic sarcophagi.

No grave robbers ventured to set foot in the swamps in the delta of Rio Tomolā. (To this day the region is so inaccessible that the Mexican government had the greatest part of the La Venta finds removed and set up again in an archeological park at Villahermosa). The excavators found the skeletons reduced to dust, of course, but the burial offerings were completely preserved. These were rich storehouses filled mostly with jade creations: imitations of flowers, plants, animals, and human beings, ceremonial axes with blades up to 10 inches long, jaguar and human masks, earrings, and other ornamental objects and decorations. Everything is fashioned with artistic perfection from this bright green, shimmering, smooth, and translucent mineral so precious to the Olmecs. They were the first to introduce jade processing in America. This does not mean, however, that the art of jade carving then developed from a simple to higher stage. Even the oldest pieces show that the artists perfectly mastered the technique of working jade. Indeed the newer finds exhibit lesser perfection than the old ones.

Why were the Olmecs so enamored of jade? This is not a natural preoccupation, for jade quarries are nowhere to be found in Mexico. Only boulders with jade cores can come into consideration as a raw material source, and how the Olmecs perceived this so improbable material in that form defies explanation. Wherever they may have come from, they must have been very familiar with jade in their original homeland. From the time of the Olmecs on, jade was highly prized by all the subsequent cultures of Central America, as clearly shown by the Olmecoid influences among the Totonacs, the Zapotecs, and the Mayas. In particular they introduced the cult of the jaguar, writing, and the calendar system which then attained its highest perfection with the Mayas. We may altogether assume, without pushing speculation too far, that many Olmec characteristics are found among the better preserved remains of the chronologically later cultures of Mexico which are not known from direct finds in the Olmec region.

Olmec ceramics preserved in Tres Zapotes are striking in the richness of forms, their manifold ornamentation and the advanced techniques used. Finally, one additional point should be mentioned: later cultures tell us that the Olmecs wore clothing that was made from tree fiber.

The unique culture of the Olmecs that emerged abruptly on the American continent exhibits many other specific characteristics, but these are most prominent: realistic sculptures, especially of human beings, quadratic pyramids with staircases attached, monolithic sarcophagi, highly developed jade processing, the first writings, the oldest calendar, the working of tree fiber, ceramics of various types and configurations.

The third region of America of interest to us includes the Andean highland in northern Peru and extends as far as Ecuador. As has been mentioned a civilization emerged in this region about a thousand years before the beginning of the Christian era, that could hardly have stemmed from the primitive aborigines of the area at first in the highlands and later in the coastal settlements. The oldest epoch bears the name Chavín, whereas the newer epochs are called Gallinazo and Mochicha. All three periods are distinguished from each other, but they also possess many common features. Characteristic of all three is the fact that new skills appear suddenly in more or less perfected form. Again these abrupt advances of civilization—as well as the very emergence of the Chavin civilization—can be explained only by the existence of outside influences.

One of the common features among the people of the Andean highlands is their remarkable habit of chewing the leaves of the coca plant. They do not chew the leaves alone, however, but mix them together with lime and other substances. Today we know why they added lime to the inside of the rolled coca leaf: this combination releases the alkaloids (cocain, truxillin) contained in the coca leaves which can then be absorbed by the body through saliva and through the mucuous membranes. The inhabitants of the Andes are so addicted to this refreshment and tonic restoration that without it they cannot perform hard physical labor. How did the people who lived here almost more than three thousand years ago know that lime had to be mixed with the leaves in order to achieve the desired tonic effect? This kind of idea does not suggest itself, no more than does the idea of adorning the teeth with gold coating, a custom that was already known in the seventh century B.C.

According to all evidence available these Andean cultures did not know writing, in any event not in the conventional sense. But they did possess a method for recording numbers and perhaps events as well without having to depend on their memory: the *quipu* cord. Thin cords were fastened to a long main cord and knotted at differently spaced intervals. The knots were of varying sizes and sometimes an additional colored thread ran through them suggesting, like the individual cords, a color code for the kinds of things that were counted. Nobody has yet been able to decipher this knot-writing; only parts that refer to numerical data have been decoded. We do not even know with certainty whether it is an alphabet or whether the cords took over the task that is accomplished today by a hand adding machine. The *quipu,* a unique contrivance in America, were developed to a highest degree under the Incas, but they are already detectable in many earlier cultures.

As they have in all cultures, archeologists also found musical instruments here. Among them is one that is particularly striking, the Panpipe. True, the idea of binding together individual cane pipes tuned according to their lengths is an obvious one. But the Mochicas blew on an instrument that consisted of two Panpipes bound together with a long cord. These wooden instruments have not been preserved but are known from illustrations. In fact, the Mochicas have left us many depictions of their daily life on ceramics. At times these representations were so detailed that two researchers, Nevermann and Bird, were able to reconstruct the related weaving techniques exactly. Textile remains from that time have also been preserved since the inhabitants of Peru left extensive mounds of debris which today constitute a fruitful source of information for archeologists. We have already mentioned that they colored cloth by elaborate methods of *plangi*, *ikat*, and *batik*. In *ikat* the threads are coated with wax or covered by another material so when the dyeing operation begins only the uncoated sections take the dye. The protective layers are then removed, and the thread is woven. There is also evidence for this unusual technique from the outset.

The Chavin living on the coast fished from seaworthy rafts, which were steered with the help of centerboards and driven by sail. Thor Heyerdahl's "Kontiki" demonstrated the seaworthiness of these rafts. There are also many accounts by seafarers of the sixteenth and seventeeth centuries who encountered these simple constructions far out on the open sea. The mode of construction of these craft has remained practically unchanged since the time of the coastal Chavin culture.

Even the oldest traces of the Chavin culture are accompanied by objects wrought in gold; copper was added only later. The ethnologist Heine-Geldern made an exhaustive study of the metallurgy of the Andean peoples and was able to prove that this could not have developed as an independent invention. We have already mentioned the complicated "lost-wax" process of casting metal, and research has uncovered a series of further skills which could only result from long development. Thus these peoples already had mastered the method of immersing gold objects in alum mineral waters and plant extracts in order to enhance the color of gold. In this way, the gold-copper alloys that were used primarily for ornamental pieces acquired a beautiful yellow color. Bronze followed copper from around the third to the second century B.C. It is improbable that the alloying process—whereby soft copper is changed into bronze by adding zinc—was invented in South America wholly independently of outside influences, particularly in the short time span of a few hundred years. Moreover, there are indications that the early carriers of the Andean cultures had systematically searched for zinc (a search that first ended in success in the trans-Andean region).

From bronze the inhabitants fashioned an unusual needle, the so-called disk needle: the blunt end does not terminate in an eyelet or a broad edge but in a disk which probably served as a handle. Among the bronze tools we find hand axes, pickaxes, socketaxes and shaftless mattocks which perhaps served as weapons. Among the decorative elements, which appear in bronze as well as in stone on ceramics, double spirals, bilateral animal representations, open-ended bangles with spirals at the end, plumed serpents with sickle-shaped wing stumps, dragon motifs and pearl necklace ornamentation merit mention here. Nor should we forget the animal figures that carry cylindrical vessels on their backs.

There are sagas and legends, some of them long preserved in writing, still known today by the populace. Might these be analyzed in order to learn where the individual cultures came from? In the case of the Aztecs this kind of an analysis seems to have been worthwhile. A number of distinguished ethnologists have perceived an historical kernel in the legend of the White God, enabling them to explain many events of this epoch that otherwise defy comprehension. But we cannot expect such results with respect to the cultures of the Central Andes and the Olmecoid peoples; too much time has elapsed in the meanwhile. The golden age of the Aztecs began around 1400 A.D., while the civilization of the Olmecs flourished fifteen hundred years earlier and that of the Chavin twenty-five hundred years earlier.

We have pointed out how improbable it would have been for these civilizations of the Northwest Coast, in Mexico, and in the Andean highland to have been formed from the aboriginal population of the American continent. Some striking features of these cultures were then described to facilitate the search for outside impulses. Let us now embark on the search proper without any great expectations. It is highly doubtful that we will find a civilization that matches our description precisely. It is well known among archeologists that in the process of cultural transmission some elements are lost, for reasons unclear to us today, while others are subject to transformation. Finally, it should be remembered that knowledge of the mother culture may also be incomplete. Let us not set our sights too high.

One hypothesis that obviously suggests itself is that members of advanced civilizations emigrated from Asia across old migration route, the Bering Strait, or along the chain of the Aleutian Islands. For a long time the Hollander Johannes de Laet (1643) vigorously espoused the view that the Scythians set out along this path. There are many arguments against this. For one thing, since about 3000 B.C. this sea ;route has been 52 miles wide at its narrowest spot, an obstacle that could hardly be overcome by a migrating people. The same applies to the Aleutians. Indeed here the stretches of open sea that must be traversed are even more sizeable. But even if the representatives of an advanced civilization could have negotiated the trek from Asia to America, they would have never managed to make their way to Peru through the by then settled regions of North America. Nor could traces of such a migratory wave have escaped archeologists. Only in respect to the northwest culture might this possibility be considered. On the other hand anyone traversing the Bering Strait would follow the Yukon and so not arrive at the southern coasts of Alaska.

If we exclude the land route, that leaves the ocean. Only those living on or near the Pacific coast come into consideration for the northwest culture or the Chavin civilization. In the case of the Olmecs there may have been an influence from across the Atlantic, but the geographical location of the Isthmus of Tehuantepec does not exclude the possibility of diffusion from the Pacific either. In fact Olmecoid traces have been found on the Pacific coasts. Archeologists have neglected this side of the land bridge between North and South America in the past but recent intensive digging in Guatamala has already borne fruit: in 1970 heads and full figures were excavated at Monto Alto in southern Guatamala, figures that are very similar to those from La Venta.

In Europe the question regarding the point of origin of the early American civilizations is as old as knowledge of them; knowledge of the special position of the Northwest Coast Indians has been fairly recent. Fancy ran riot as long as they were not sufficiently investigated and objective methods of dating were not available. Thus Bartolomé de Las Casas believed to have recognized in them the descendants of the lost Ten Tribes of Israel.

Then the exponents of the Atlantis hypothesis had the floor. If the inhabitants of this sunken continent could be made responsible for all kinds of historical absurdities, why not also make them responsible for the advanced civilizations of America? An insurmountable obstacle soon placed itself in the way: the date. Plato, whose account of Atlantis is the basis for all conjecture, dates the submerged kingdom around the year 6500 B.C. At this time not even the art of pottery-making was known in America, the first stage of advanced civilization having appeared in the tenth century B.C. with the Chavin. A similar continent, which also sank to the bottom of the sea, but which first sent its inhabitants in all directions of the compass, is the continent called Mu, its existence postulated by Churchward. In contrast to Atlantis, however, Mu was supposed to have been located in the Pacific.

The views of those who contend that the impulse came from the Mediterranean area deserve greater attention. In 1970 Thor Heyerdahl, in a bulrush boat built according to old Egyptian models, succeeded in sailing across the Atlantic to America. His first boat (1969) had fallen apart (Heyerdahl had negelected a seemingly unimportant detail during its construction). On a similar route Cretans or Phoenicians might have reached Middle America in prehistoric times. It seems clear that the fleets of the Cretans went through the Pillars of Hercules (the Straits of Gibraltar) into the open Atlantic. The Phoenicians had sailed around Africa; why should they not have succeeded in making a voyage directly across the Atlantic?

We can eliminate the Vikings immediately, although opposing views exist. They did discover the northeastern coasts of America, but by then the kingdoms of the Olmecs and the Chavins had long disappeared.

The Pacific remains. Here there are only a few promising candidates for sea voyages of this kind. The inhabitants of Polynesia might easily have made a Pacific crossing. But they could not have brought an advanced culture. They also would have come too late, since the islands of Polynesia were first settled in the post-Christian era. Japan, China and the area of Indochina are more likely candidates. Of China particularly, we know that junks undertook extensive voyages across the open sea. By the fourth century sea voyages from China to Java no longer seemed like a foolhardy adventure. Earlier accounts also report that merchantmen from China passed beyond the Straits of Malacca and called at ports on the southern tip of India directly across the Bay of Bengal. Did they also make a successful voyage across the Pacific? That they possessed the necessary cultural level to influence the Olmec and Chavin civilizations is beyond question.

There is yet a third hypothesis: in his book *Men out of Asia* (New York, 1947) Harold Gladwin makes the fleet of Alexander the Great responsible for the advanced cultures of Central America. When Alexander suddenly died in 323 B.C. the fleet that had been assembled was left leaderless. It disappeared from history's field of vision, according to Gladwin's hypothesis, because it sailed towards India and reached Central America after a long voyage with many intermediate stopovers. He also contends that the first settlement of America took place in several waves across the Bering Strait. His theory has never been accorded the attention it deserves, perhaps because he clothes it a witty and flippant style and pitilessly impales isolationists on his pen.

Once more, the impetus for the development of advanced civilizations in Mexico and the Andes highland must have come from members of an already existing advanced civilization. Experts have placed the beginning of the Northwest Coast culture around 1000 B.C., in which case the impulse must have come from an area whose inhabitants even then could look back upon a long history. On the basis of the explosive leaps of development in the Central Andes region we must infer a series of impulses stretching over a period of centuries. The Olmecs began around the first year of the Christian era with a new thrust of influence.

The Mediterranean peoples definitely meet these criteria. The dynasties of Egypt stretch as far back as 3000 B.C., the great pyramids of Cheops and Chephren were already in existence in 2500 B.C. Even after Alexander interrupted Egypt's independent development in 332 B.C., the advanced level of civilization was preserved, albeit under Greek, (and later, Roman) domination. The same can be said for Crete, even though the Cretan-Mycenaen culture was destroyed by seafaring peoples around 1200 B.C. A reflection of its earlier greatness, however, subsisted under Greek influence. Nor are the Phoenicians to be excluded *a priori*; after all we find their colonies everywhere on the shores of the Mediterranean and also on the Atlantic coast.

Whether they would have passed on the knowledge of a far-advanced civilization is questionable. Their settlements bear more the character of trading posts than of cities where art and culture flourished.

From the Pacific side, Japan may also be taken into consideration. The first pre-Christian millennium was marked by the Yayoi period, which worked bronze, yet continued to use stone implements for a long time. Not until the middle of the fourth century B.C. did a greater empire emerge from the countless small states. Japan might have had occasional contact with Central America, but would hardly have maintained a more or less regular, long-term relationship. Only a huge empire with a thousand-year long continuity would play this role. This points to China. If we disregard the legendary Hia kingdom, we find a superbly organized State in the region of the Yellow River up to the Yangtze Kiang, the kingdom of the Shang dynasty (1520 to 1030 B.C.). During the subsequent period of the Early Chou (to 722 B.C.) the empire extended far north and south and also included the greatest part of the coast of the Yellow Sea. Even after the reign of the Early Chou collapsed and the empire broke up into independent, warring States, the cultural level that had already been reached subsisted unimpaired. In 221 B.C. Shih Huang Ti created the first Chinese great empire, and the Han dynasty (202 B.C. to 220 A.D.) ushered in its golden age. Here was a civilization whose continuity can be evidenced since 1500 B.C. at the latest, and one that could most probably maintain contacts over enormous distances.

But what about Alexander's fleet? Even in the improbable case that they made the trip from the mouth of the Indus up to Central America at that date, the voyagers would have arrived too late to found the Chavin culture. It is true that they could have influenced later epochs, as well as the Indians of the Northwest Coast. Nevertheless here we wish to exclude the fleet from further consideration. Many of the influences that warranted Gladwin's postulation of this truly astounding voyage can be explained far more simply in the light of recent research.

The Egyptian, Cretan, and Chinese or–more accurately–East Asian spheres of culture have met the first requirements in our search. Who among them possessed the technical means to carry out a long ocean voyage of such a kind?

The Egyptians undertook sea voyages to very distant regions as early as the second millennium B.C. To be sure, we still do not know exactly where the "divine land" Punt, the destination of regular expeditions, lay. But it seems to be certain that a fleet sent out by Pharoah Thutmosis III, around 1490 B.C., circumnavigated the African continent. Various signs indicate that the Canary Islands were likewise visited by Egyptian ships around 1250 B.C. From there the northeast trade wind would have propelled a sailing ship across the Atlantic up to Central America in a relatively short time. The boats of the Egyptians, were up to 150 feet in length and equipped with oars and sails, would have certainly stood up well during this voyage. This was proven by Thor Heyerdahl's last sea voyage.

As far as ship-building is concerned, a trip of this kind could also be attributed to the Cretes. For several centuries their fleet dominated the Mediterranean. Its sway was so undisputed that the palace of Knossos required no walls whatsoever as protection against attackers. Anyone who wanted to approach the isle of Crete first had to pass the aquatic border post. Cretan ships sailed on the Atlantic and thereby may have discovered the Azores. This was already a goodly stretch in the direction of America, but the Azores are little suited as a jumping-off point for a further thrust: the prevailing wind and currents put sailing vessels at a great disadvantage. Nevertheless we do not wish fully to reject the idea that Cretan or Phoenecian seafarers succeeded in making the journey across the Atlantic. (This hypothesis is treated very thoroughly by P. Honore in his book *In Quest of the White God).*

If we now turn our attention to the Pacific, someone is bound to object. However, despite the fact that a voyage from East Asia to America is at least twice as long as one directly across the Atlantic, the Pacific ocean offers very good conditions for a voyage of this kind. The prevailing winds and ocean currents actually favor sailing vessels. A ship starting out in the Chinese Sea can count on westerly winds between 35° and 55° latitude; they blow the whole year round. At the same time the north Pacific current would further accelerate the voyage. These favorable conditions constantly prevail whether the ship forges ahead northward along the coast, or whether it steers into the open ocean. It is known that long before the Christian era, Chinese ships sailed past Formosa and Japan. They surely were aware of the prevailing wind and current conditions.

Upon arriving on the American side, the Asiatic ships could easily continue the voyage southwards within sight of the coast. The northeast trade wind and the equatorial current are effective navigational aids for the return journey.

On a circular tour of this kind it would be difficult to discover the Hawaiian Islands since on the journey outward they lie too far south and on the return journey too far north. Two hundred years ago Spanish ships sailed along the route that we have just described from Mexico to the Phillipines and back. And because the Spanish mariners found the optimal course, the existence of Hawaii remained hidden to them.

At this point the reader who has looked at a map of the Pacific may have some questions. Would ships on this course tend to wander into the maze of Polynesia's islands? In any event, traffic of this kind surely would have left traces behind among the people dwelling there. If a long-term contact between East Asia and America ever existed, why not look for remains on these islands? At first sight this seems like an attractive idea; but in reality the many Polynesian islands are entirely lost in the enormous expanse of the Pacific. Even if a mariner finds himself right in the middle of them, he sights land only very seldom. When Magellan, coming from Tierra del Fuego, steered towards the Philippines, his ship sailed right through this island world. And although the crow's nest was constantly occupied the shout "Land ahead!" resounded only twice during the three-month voyage. The first time the lookout sighted two bleak, uninhabited islands, the second time the Mariana Islands. Thus it signifies nothing at all if no traces of trans-Pacific voyagers are found on these islands between Asia and America.

There is even another indication that wind and weather favored a connection between both coasts of the Pacific: Asiatic ships were repeatedly driven by storms up to the shores of America. The California Academy of Sciences, compiling a list of all occurrences of this kind, found that between 1800 and 1950 alone more than fifty ships, partly Japanese, partly Chinese junks, were driven onto the coast of California. In most cases the crews withstood this involuntary ocean crossing very well. To this very day beachcombers constantly find large glass balls on the north American Pacific coasts. Most of them have no idea of the long journey these balls have made. They are the buoys of Japanese fishing nets that have broken off from their casings and have been carried away.

A find that was made only a few years ago merits mention here in connection with involuntary voyages: a group of archeologists discovered near Valdivia in Ecuador a very old settlement quite distinct from all the others that had been excavated in that area. The pottery found there led scientists to believe that the locality was founded by Japanese fishermen of the middle Yomon culture. The ornaments excavated are practically identical with those which have also been found on the southernmost of the Japanese main islands, on Kyushu. If this interpretation is correct, we could have in Valdivia a Japanese enclave that came into being in the third millenium B.C. Here it can have been only a case where fishermen were thrown off course by wind and currents. The existence of a settlement of this kind would resolve one of the (many) riddles of South America, for up to now it has not been known whence the very early South American civilizations derived their knowledge of pottery-making. All indications refute the notion that this knowledge came with the last migratory wave from the north, and the possibility that it was an independent invention is also slight. Where, then, do these bowls and jugs stem from? Perhaps the Japanese shipwrecked on the coast of Ecuador taught pottery-making to the aborigines. This hypothesis would fit in terms of dates: simply fired pottery wares appeared in South America around 3000 B.C. Further traces that point to Japan have not yet been found in Valdivia.

The external conditions for contacts between the Far East and America were favorable. Unfortunately, not much is known about the sea craft of the Far East. There is no doubt that the coastal inhabitants of China and Indochina possessed large double canoes very similar to those used by the Northwest Coast Indians. Under the command of experienced seafarers such boats are capable of crossing the Pacific. We know that the inhabitants of Polynesia undertook voyages of several thousand sea miles in their considerably smaller outrigger canoes. Thus, for example, the 2000 sea miles from Tahiti to Hawaii were by no means considered a major undertaking. The double canoes of East Asia, which were up to 75 feet long, were even better suited for such a trip. They could accomodate about 50 men and store the necessary supplies and equipment on the platform that connected the two boats.

In contrast to these canoes, each one carved from a tree trunk, there were also plank-sewn boats on the coast of Asia after around 900 B.C. Their remains have been found on Formosa and in the Phillipines. Indeed, it seems as though they might be the same ships that as late as the last century known as "Korra-Korra," conducted trade off the Maluku Islands. With two masts and a length of as much as 90 feet these were quite estimable sea craft. Their seaworthiness was often increased by one or two centerboards, an indispensable device for long ocean voyages; otherwise they could sail only before the wind.

Still a third seagoing craft known in Asia could cross the Pacific: the sailing raft. Like the coastal dwellers of South America, those who lived along the sea in the region of present-day Vietnam and south China also built these clumsy-looking contraptions. These craft tacked about off the coasts of South America in great number as late as the sixteenth century. Bartolomē Ruiz de Estrada, the pilot of Cortez, has reported on his encounter on the high seas with a raft of this kind. The

Indians aboard were on one of their regular trade voyages to the North and were carrying gold, silver, wool, blankets, mirrors, dyes and feathers as cargo.

Not only were the natural conditions for voyages to America, starting out from Asia, suitable, but human beings also had at their disposal the means to exploit these advantages. Whether they conquered the Pacific in double canoes, plank boats, or rafts is a question of secondary importance. If they really accomplished this extraordinary seafaring feat–and this not only one–we must ask what motives they had for setting out on such a long journey. Was it lust for adventure, or were the seafarers in flight from an enemy? We know that Emperor Shih Huang Ti, in 300 B.C., sent out several expeditions that were to bring him back intelligence about the Islands of the Immortals. Voyages of such a kind had already taken place long before his time, and many expeditions must have even returned since legends tell of light-skinned people who lived far away on the other side of the ocean. This could not have been a reference to Japan, since that country was well known and was many years too close in terms of the duration of the voyages as described by legend.

Heine-Geldern offers another motivation, already mentioned here: the Pontic migration. In the second half of the ninth century B.C. peoples broke out of the region around the Black Sea and set out in an easterly direction. Among them were found Illyrians, Thracians, Cinmerians, Caucasian tribes, Tocharians, those who lived in the immediate area of the Mediterranean, and Scandinavian and Germanic tribes. Why they set out on this migration is not known, but we can follow their route up to China; in 771 B.C. the kingdom of Western Chou collapsed under their assault. This massive migration first came to an end at the Pacific coast, between the Gulf of Tonkin and the mouth of the Yellow River. Is it possible that some groups pressed forth even farther, to the water, and finally reached America? Or, perhaps, were there people who were fleeing before this migration of peoples?

There are a number of arguments against the hypothesis of a

trans-Pacific contact, but no final evidence has been presented to indicate that no contacts took place in prehistoric time. The plough, the cart, the potter's wheel, stringed instruments, and bellows no doubt existed in the countries on the west coast of the Pacific in 1000 B.C. while nothing of this was known in America before Columbus landed there. Nevertheless it is well known that when a transmission between two cultures occurs all elements are by no means carried over. Again and again during such encounters, for reasons now unknown, specific single skills were not transmitted. One might well ask whether the cart would have been of much actual usefulness in the mountains of the Andes; for the wheel was known in both South and Central America long before the beginning of the Christian era. It did not serve to transport loads, however, but appeared exclusively as a children's toy. Thus archeologists have found clay animals as burial offerings which were fastened to axes with cylindrical wheels and which could be drawn on a string.

Another case presents an explanation that is even convincing: in the very early stages of Chavin Civilization gold was the only metal known. If (as we assume) this central Andean culture was influenced by China copper and bronze could not have been missing. But what would actually happen, say, if a relatively small group were driven off course, driven ashore on an unknown coast? Its members would certainly begin to look around for copper or suitable ores in order to fashion tools out of them in the customary way. If this search is not successful within the time span of the first generation, the knowledge is lost, and the settlers must content themselves with stone implements (which were always used in bronze cultures). The search for metal will not be resumed until new groups arrive. This seems to have been the case in the Andes highland: not until three to four centuries after the working in gold do we find there traces of copper, and later, of a bronze industry. This indicates manifold contacts, particularly since the copper articles which appear so suddenly attest to a perfect mastery of technique.

Viewed as a whole, however, we cannot satisfactorily explain why certain cultural elements, among them the eminently useful plough and potter's wheel, were not transmitted. But why, for example, were the inhabitants of Polynesia still living in the Stone Age when they were discovered by Europeans? They surely came from the Indochina region at a time when knowledge of metal was generally diffused there. They seem to have "forgotten" how to work metal after their migration to the island-world of the Pacific. We can come to an analogous conclusion with respect to the East Asian influence on the civilizations of America. Amid the many achievements that were transmitted, the plough, cart, potter's wheel, bellows, and stringed instruments were lost in the process.

Upon closer scrutiny, the arguments of the isolationists presented earlier prove not nearly as clear-cut as they seemed to be at first. Although we cannot refute them all, let us continue to pursue the theory that the cultural leaps observed in America were inspired by people from the opposite shores of the Pacific. Many hardly contestable proofs can be mustered to buttress this thesis. One item involves the plants that man has cultivated for nourishment. If the same domestic plants are found in Asia and in America, their point of origin must be determined. The following will try to show that some plants found their way from Asia to America and that this could happen only by human transport. The seeds of these plants were not brought in by the migratory waves that came across the Bering land bridge, however. Either those plants had not yet been cultivated or they could not have withstood a long migration of that kind through unfavorable climate zones. Nor is it plausible to contend that the same culture-plants were cultivated by people on the both sides; that would really be an absurd coincidence.

The first point to prove is whether the plants under discussion really were already in America before Columbus. In each case the question arises whether they might not have been brought along by the Spaniards, and frequently the answer is not so easy to provide. The coconut palm is an example of this.

It cannot be imagined as absent from daily life in southeast Asia; its existence in those regions before the first millennium B.C. has been ascertained beyond any doubt. But we also find the coconut palm in droves in Central and South America. Did it grow there before Columbus? The heated controversy over this point died down only when a letter written by Alvarõ de Guiyo to Cortez became known. In it de Guiyo promised to send his captain the nuts of the coconut palm from Panama—a gift which in 1539 was costly enough to warrant mention. Thus the Spaniards did not introduce this many-faceted fruit; it must been have come to America from its homeland in Asia. But how? Did it set out from the Asiatic mainland slowly conquering one island after another, aided and abetted by the ocean currents, and finally reaching America from the easternmost Polynesian islands? Northerly and southerly equatorial countercurrents would have propelled the coconuts in the desired direction. It can easily be calculated that a voyage from the Marquesas Islands to Peru would take around seven months. A coconut, however, cannot float in water for such a long time. Experiments have shown that the husk gradually sponges up sea-water. After three or four months the fruit loses its buoyancy and sinks. Thor Heyerdahl stored a number of ripe coconuts under his raft in order to determine whether they would survive the voyage. Not a single one of them germinated when he planted them after the three-month voyage. On the other hand, he was more successful with those that he had brought along in baskets on the deck of the raft. They had come in contact with the ocean water only occasionally; almost half of those planted sprouted shoots.

It is clear from the foregoing that the coconut palm did not reach America from Asia along a natural route. It could have withstood the long journey across the open sea only in man-made sea vessles. Durable and tasty, the coconut is superbly suited as a provision for long journeys and was doubtlessly found in any boat that dared the trip across parts of the Pacfic. Upon arriving in South or Central America in a form capable of germination, it found ideal conditions for growth and survival and spread rapidly. Just when this happened is unclear, however.

Nor do we know when the gourd or calabash (*Lagenaria Siceraria*) reached America from its Indian homeland. Its presence in Peru dates back at least 2500 B.C.; it has been known in the southern United States since 400 B.C. Neither the gourd itself nor its seeds can withstand long periods in salt water. Even if it should happen to survive, the plant could not take root on a beach. Indeed, very few among those plants or plant seeds that were washed ashore would find suitable life-conditions there. Most would die. Thus the gourd also can only have arrived in America only by human carriage.

The history of the batata–the sweet potato–is especially interesting. Its homeland is Central America, yet already long before the first Europeans it reached Polynesia and from there, Asia. How is that to be explained? This tuber cannot survive even a brief trip in salt water, and since it propogates through offshoots no diffusion by ocean current is possible. The only remaining possibility is that the inhabitants of Polynesia reached the coasts of Central America and took sweet potatoes on board on the return voyage. An indication of this is that the Polynesian name for the batata, *kumar,* is also the name for the sweet potato in one part of Central America. That voyage must have been a remarkable bit of seacraft. The outward journey perhaps was less so since an easterly course starting out from the Marquesas Islands after some 3500 sea miles leads to a landing point in America. The navigational accomplishment necessary on the return voyage is considerably more noteworthy. The islands form a mere speck in the expanse of the Pacific, and it is extremely easy to sail right by them. This journey (or these journeys) must have taken place in the distant past since in Polynesia there are many cultivated variants of the batata. The tubers play an important role in magic rites and ceremonies, which always indicates a long history of possession.

Research has concerned itself much more with the problem of cotton. Where does the plant come from? There is no doubt that it was known by the oldest Asiatic cultures; thus we already find it in 3000 B.C. in Mohenjodero in the Indus valley. But it has also been in America for a very long time. The problem is complicated still further by the fact that several wild forms exist in America. Researchers, however, were quite taken aback when microscopic investigation disclosed that American cotton contained two sets of genes, one set of the American wild form and the other of the Asiatic cultured variant. Scientists are not in full agreement, but the majority champion an Asiatic homeland and later transmission to America (long before the beginning of the Christian era, of course) with subsequent cross-breeding.

The origin of another plant formerly thought to be typically American–corn–is still disputed. Corn seems always to have existed in Assam and Central Asia. The same basic plant appeared in Peru long before the Spaniards. Among some Indian peoples–the Aztecs, for example– legends tell how corn was brought to them as a gift of the gods in remote antiquity. This event is still celebrated in feasts. Might this be a reference to visitors from the Asiatic coasts? Here, too, no clear answer can be made.

There is still a further indication that many plants reached America by transport across the Pacific. If we set out to determine the number of botanical species on the Pacific Islands, it appears that the number decreases as we move east. This indicates the Asiatic origin of the flora. Most of these varieties, if not all, could have been brought there only by human beings. It seems logical to assume that some plants will have completed the Pacific crossing and found a new homeland in America.

In the case of domestic animals there are fewer indices that point to an Asiatic origin. Only the pig, dog, chicken, and rat reached America from the Asiatic coastal areas of the Pacific before Columbus. In fact it was long disputed whether chickens existed in America before the Spaniards, but this was finally proven. The original bird was a tailless breed from South America that laid blue eggs. A tailless breed of chicken is very rare, so it was not long before its original homeland was found: Japan.

Like the plants, these breeds also could have crossed the Pacific only in the company of human beings. Pigs and chickens surely served the early travelers to America as fare, the dog was a fellow traveler and the rat a stowaway. That only these four animal species arrived in America before Columbus is another indication that contacts were made along sea routes; had there been a land bridge in historical times, the horse, donkey, sheep, goat, and ox would have arrived in America early.

Two special hunting methods may be mentioned here, both so unusual that they could hardly have originated independently of each other. Yet they were practiced in Asia (China, Japan) and America (Mexico, Peru). The first method is catching fish with the cormorant. This voracious creature, classified among the web-footed sea birds, is found almost throughout the world and feeds on fish which it hunts and catches under water. The cormorant is an extraordinarily skillful hunter and therefore can meet all his daily nutritional needs in a half hour. This has been exploited by men in Japan, China and Peru, who capture young cormorants and train them in such a way that the grown bird brings the caught fish to the boat. The owner then opens the beak, retrieves the still squirming fish from the bird's gullet and sends the cormorant out after more prey. In order to prevent the bird from swallowing the fish in the excitement of the hunt, it is made to wear a ring around its neck so tight that only tiny fish can pass through its gullet. When the day's work is over the ring is removed from the neck of the bird and it is permitted to hunt for a brief time. This fishing technique could have hardly developed on both Pacific coasts independently from each other. Since the Spaniards sent reports about it from Peru it is also an indication of transpacific contacts long before Columbus. But the second hunting method is even more striking.

Any Mexican who lived near a pond where wild ducks came could vary his domestic menu with an occasional roast. At that time there were no guns, however, and a method had to be found that would not disturb the ducks. He found some gourds as big as a man's head, hollowed them out, closed them again, and brought them to the pond, where they floated on the surface of the water. Soon the ducks grew used to the drifting objects and no longer paid any attention to them. This was the psychological moment for the hunt: the man hollowed out another gourd, but cut holes in it for his eyes and nose. Wading into the water in the ducks' absence, he fitted the gourd over his head. The returning ducks, finding only floating gourds in the pond, had no presentiment of peril, not even when one of the gourds moved slowly towards a lone duck who was resting to one side of the pond. In a flash, the duck was seized by its two feet and pulled under the water. No desperate cry, no anguished fluttering of wings; and that night roast duck was served on the Mexican's table.

Substitute "Chinese" for "Mexican," and the story that we have just related would still be true. Who brought that knowledge to the inhabitants of Mexico, and particularly to those in the swampy coastal regions such as the Isthmus of Tehuantepec? An inhabitant of East Asia did, and he was there long before Columbus left Spain with his three caravels.

Thus far we have spoken only very generally of trans-Pacific contacts, but the following passages will show where and when these contacts originated. Let us turn first to the central Andean culture since the many finds that have been made there can most easily be fitted into a uniform picture.

The stone beast of prey carrying a cylindrical vessel on its back stems from the oldest period of Andean culture. (See figure below.) The combination is certainly not one that obviously suggests itself, nor did the artist ever see one in nature. Did he perhaps have some prototype for it? A sculpture has been preserved from another culture that may be considered as the prototype for the Andean stone: it is a feline of bronze bearing a cylindrical vessel on its back. This figure stems from China, and dates back to time of the Chou dynasty, between the ninth and eighth centuries B.C. This was a time

Left: bronze tiger from the Chou Dynasty, China. Right: stone beast of prey from the Chavin Culture, Peru. The two sculptures must have been stamped by similar culture areas. Also note the double spiral projecting into a bird head on the side of the stone figure. This motif is found elsewhere only in China.

when China was only nominally ruled by the Chou, while the individual kingdoms enjoyed great independence. Can the congruity of the two sculptures in China and South America be coincidence? Not only did they originate at practically the same time, but the ribbon-shaped double spiral on the side of the figure from Chavin de Huantar is likewise a motif of the Chou period. In addition, the spiral projects into a bird-head, which is also a motif from the Chou period. Here we must assume an influence exercised on the early Andean cultures by representatives of the Chou period from distant China, for the ribbon-like double spiral with the bird head thus far has not been seen anywhere else.

The Chavin sculpture is by no means the only hint of Chinese influence. For example, a bas-relief that also depicts a feline predator stems from the same time (Chavin). Again we find the double spiral stemming from China, again the spirals project into a bird head. The tail of the bas-relief is also noteworthy: it is scaled. This becomes understandable when we know that in China the scaled tail was also represented on animals that had nothing to do with water. Still other ornamental motifs of a purely Chinese character have been found in the Andean highlands, such as the intertwined, coiled fabulous creatures with the sickle-shaped dragon wings of the Chou style. It is impossible to explain this away as coincidence. The hypothesis of a transpacific cultural relationship is a much more reasonable explanation.

Along with the emergence of artistic motifs from the middle Chou period of China, we come across a technique of working gold in the Chavin culture that seems to appear out of the blue. It is difficult to make the necessary comparison with works of the Chou period wrought in gold, since few works fashioned from precious metals have been preserved from the China of that time. Nevertheless a relationship can be derived here from the hammering technique. The paucity of gold finds is simple to explain: old works wrought in gold were repeatedly melted down and worked into new, more modern creations. In some Asiatic cultures, in fact, gold first became a thing of value when the gold piece was redesigned and then blessed by a priest. Foreign conquerors, grave-robbers, and tribute-paying subject states all demanded gold. How many very ancient works wrought in gold from the Chavin culture must there have been that were melted down in Peru on Pizarro's orders?

But why did the oldest Chavin culture know nothing about bronze? If artistic motifs and techniques for working gold were already transmitted from China, why not the methods of making bronze? Bronze did arrive relatively late in China (at the end of the Shang dynasty), yet bronze-casting achieved a high point there which was never surpassed in the whole world. Moreover, it must be recalled that the introduction of bronze by no means signified that stone implements were immediately discarded. Even at the time of the Han dynasty–one thousand years after the first bronze castings–many stone implements were used. Bronze was expensive, and the tools of the early Stone Age were not all too inferior to those fashioned in bronze. Another point must also be taken into consideration in the case of the Andes cultures: bronze casting requires–in addition to the raw materials copper and zinc–through technical knowledge that could be obtained only through personal experience. If the first arrivals from China found neither copper nor copper ore in their new domicile, the subsequent generation would still know about copper but would no longer be able to make it, having never had the opportunity to pick up the necessary practical knowledge. This holds true doubly for bronze. In addition, it must be noted that the second component of bronze, zinc, is not found in the central Andean region.

The absence of copper processing in the Chavin culture supports the argument that the transpacific contacts were by no means single, once-and-for-all events. At the end of the Chavin epoch, in the sixth to the fifth centuries B.C. (the beginning of the Gallinazzo culture) a highly developed technique of working copper originated suddenly in northern Peru. This was not the beginning of primitive copper processing, but already involved techniques as complicated as the "lost-wax" process, technique so difficult that the peoples of Mesopotamia required two thousand years of steady development to master this casting method. Yet in the central Andes region it simply emerged, as if someone were to build a television set without knowing anything about electricity. Again, someone from the outside must have come who possessed this knowledge and transmitted it. Nor could it have been a brief visit. Rather, we must assume a long sojourn or even a permanent settlement.

For a long time, approximately up to the first century A.D. we find in the Andes a copper—but no bronze—culture. This, however, is due only to the fact that no zinc deposit could be found. There are indications that several expeditions were sent out which pushed forward to Bolivia and Argentina. There, zinc was found later on the eastern slopes of the Andes. The Gallinazo and the early Mochicha epoch were rightly called copper bronze-cultures.

Still another characteristic of the Andean cultures probably came out of China: the knotted cord, known as *quipu* in Peru. Knotted cords were still being used at the turn of the century by primitive tribes on Formosa to represent figures, amounts and perhaps even news. This was surely a relic from very ancient times for this aid was no longer mentioned once it was supplanted by the writing brush in the Han dynasty. But it must have been known on the mainland earlier. Thus knotted cord writing is mentioned in the *Book of Changes,* a work that originated in 300 B.C. We have, moreover, proof that it was already of great antiquity at that time. In the *Tao Te Ching*, Lao-tse writes in a tone of lamentation:

> Give the people knotted cords again
> And let them use these,
> Then they will relish their food,
> Take pleasure in their clothes,
> in their peaceful domiciles
> and in their customary chores.

This standard book of the Taoists was presumably written at the time of the Warring States (480-211 B.C.). From the text cited here we can infer that knotted cords were once used in times long gone by—at least by the beginning of 1000 B.C. when the first trans-Pacific contacts took place.

Here it should be pointed out that culture elements transmitted up to 500 B.C. are of a purely Chinese character. After this date, other influences emerge in South America which suggest that there was also contact with another culture area. This culture area exhibits Chinese traces, but also contains unmistakable elements from the region of the Caucasus. These provided the basis for Gladwin's theory—now discarded—that Alexander's fleet sailed from the mouth of the Indus up to Middle and South America. Only a few examples are named here.

As early as 1883 Virchov puzzled over the fact that the disk needles of the Caucasians are practically identical with those of the Peruvian Andes. Today there is significantly more material at hand and the congruence of the disk needles is by no means an isolated example. There are needles whose blunt ends are formed into double spirals, animal heads, or lateral eyelets; we find them in the Caucasus and in the Andes. One of them even depicts a complete hunting scene with two dogs and a stag, and we find its stylized parallel in Peru. The double-edged hand axe

On vessels painted in Peru warriors were depicted with axes like those above. (Demon figures of the Mochica Culture.)

with a projecting stump for the attachment of the shaft appears in the cultures of Peru. In the Old World we find it in Hallstatt, in Inner Mongolia and in China. Whether it is fashioned of bronze, copper, iron or stone, the kinship is unmistakable. Socket axes, pick axes and mattocks also appear on both sides of the Pacific. If only a single form from the Andean region were similar to an Old World form the parallel development could be explained by arguing that "the same function compels the same form." But this is no longer possible in the face of so many parallel finds. Other examples are the pearl-necklace motif (especially known from the time of the later Chou in China), belt buckles with filigree work and S-shaped double spirals, tweezers, star-shaped club heads, grooved needles and tiny bell pendants. The pair of bronze mirrors, one from the Ordos desert region of Inner Mongolia and one from the Andes highland (see below), serve as an especially impressive example of a transmission from the Old to the New World.

Thus far we have discussed only the common traits between metal-working techniques and objects, but the trans-Pacific transmissions hardly ended there. This is not the case. One researcher, Nevermann, has compared the weaving and dyeing techniques of southeast Asia with those of the Andean cultures. Both used the hand loom with a back strap. A vase painting has even been preserved from the Mochicha culture which depicts Indian women weaving. Woven fabrics originated in the hands of Asiatic women in the same way. Can we conclude

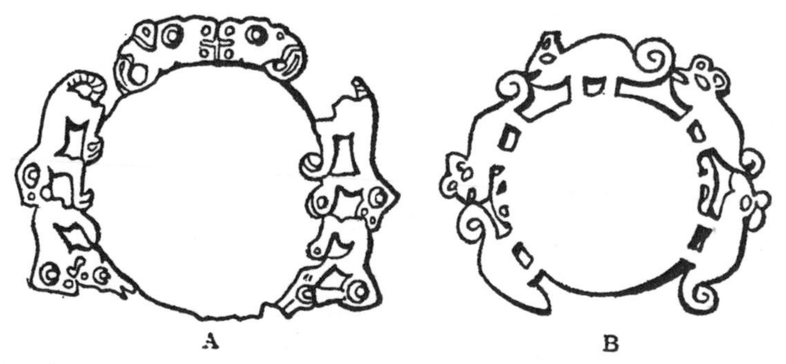

Two bronze mirrors. One stems from the Ordos region, Inner Mongolia (A), the other from Catamarca, Argentina (B). The resemblance is unmistakable.

from the existence of this technique that women also went along on the voyages clear across the Pacific? This too would suggest that these contacts were not accidental but took place in a planned way. The unusual dyeing techniques *plangi, ikat* and *batik* known throughout in Indonesia, are found again in the fabrics of the Mochicha culture. Though metallurgy and weaving are found among many peoples, the custom of plating teeth with gold is only found on the coasts of the Pacific. At the time Marco Polo traveled in China this custom was already very old in the Yün-nan province and could be traced back at least to the Han period, but was probably even of greater antiquity. The same custom was also widespread in Indonesia until recently. Heine-Geldern describes the three different decorative modes of the Batak (a Malayan tribe) as follows: "Tiny round plates of gold, copper, brass, or mother of pearl set in the teeth or gold strips set between the teeth, incisors covered with gold, silver or wire, middle part of the incisors filed down and circled with a narrow strip of gold sheet metal that is fastened to the outer incisors or the eye teeth with tiny nails, finally the teeth and, at times, the gums encased with gold sheet metal." If we wanted to describe how the dandy in the Andes beautified his teeth twenty-five hundred years ago, we could use the very same words.

Curt Sachs, after comparing the musical instruments of South America with those of Asia, came to the conclusion that here too a similarity prevailed that could not be ascribed simply to parallel development. The Panpipes may serve as a model. Just as in China and in Indonesia, every second pipe is taken out of its normal succession of notes and combined with a second Panpipe. In China as well as among a South American tribe, it is said that thus the sundered female and male tones are joined together. Both Panpipes are tied by a cord, and a vase painting from the Mochicha epoch depicts how this instrument is used.

In conclusion it should be further pointed out that in both southeast Asia and in the Andes countries natives chew leaves as stimulants. In Asia it is the leaf of the betel plant, in South

The people of the Mochica Culture in the Andes left many pottery paintings depicting scenes from their daily life. Significant inferences of transpacific influences can be drawn from these pictures. This illustration shows Indian women at the hand loom with a back strap, which was also in widespread use throughout Asia.

America that of the coca plant. Both cultures use lime in order to release the substance that produces the tonic and refreshing effect. Again we see here two features that are separated by the wide expanse of the Pacific, but which nevertheless must have a connection with each other. This impression is further strengthened if we compare the tiny can in which the lime is kept and the spatula that serves to dig it out: the congruency is unmistakable. These are by no means the only examples to be found.

Here the place of origin is rather easily identified in regard to the early Chavin culture and up to 500 B.C: the cultural impulse could have come only from China in the Chou period. Thereafter we run into difficulties. How are we to bring the cultural elements from the Caucasus, the Ordos desert region, China of the late Chou period, IndoChina and Indonesia under one head? The solution has already been mentioned: the Pontic migration. In the course of the migration of peoples, cultural elements from the Black Sea region and the Caucasus arrived in the Far East. As thorough investigations have disclosed, the stream of peoples split up into many branches on the eastern end of the Nan Shan mountains, following the natural courses suggested to them by the terrain to the east and the south. One part pushed forward along the plain of the Hwang Ho (Yellow River) and of the Yangtze valley, while another made its way directly southwards in the region of present-day Vietnam including the region of the Gulf of Tonkin. The Dongson culture then sprang into being, quickly spreading out over the whole of Indochina and the Malayan archipelago up to New Guinea. Apparently even the aborigines of New Zealand, the Maori, brought cultural elements from this area in 400 B.C. Many motifs of the Late Chou period have been clearly preserved.

Because of its relationshp to the Pontic migration the Dongson culture naturally contains influences from all the regions through which these peoples passed over several generations on their long march from the steppes between the Black and Caspian Seas up to Indochina. In addition to these were influences from the neighboring kingdoms, nominally subject to the Late Chou dynasty. When the carriers of the Dongson culture later crossed the Pacific, they brought influences from the Caucasus, Inner Mongolia, China and Indochina with them.

Unfortunately only little is known about the Dongson culture. Thus far no writings have been discovered; the excavations are still incomplete. The wars that have been raging in this region for the last thirty years have done little to further the excavation projects of the archeologists. One thing, however, seems to be certain: this culture was not linked to a single kingdom, but influenced many individual and independent regions. The core regions remained North Vietnam and the Gulf of Tonkin (where the Pacific voyages began). These regions lost their independence when China under the Late Han incorporated them, proclaimed them Chinese provinces and sinified them. At this time, around the first century A.D., connections with South America were broken off.

Transpacific contacts can likewise be assumed for the second region of America, whose history occupies a special position. The indices for this are not exactly plentiful with respect to the

Olmecs but there are several reasons why we should not expect to find any. The old Olmec culture was destroyed by foreign conquerors after a bare two centuries. Hence it had too little time to leave direct and lasting traces behind. Moreover, in the humid hot climate of the Isthmus of Tehuantepec any vestiges would rot rapidly or be covered by the jungle. In contrast the dry climate of the Andes highlands is far better suited for preserving the traces of old civilizations through the centuries.

Some elements of the oldest advanced civilization of Mexico, that of the Olmecs, bear pronounced Asiatic features. Let us recall that the Olmecs suddenly appeared around the beginning of the Christian era with a fully developed advanced level of culture. Since transatlantic connections are rather improbable, as we have seen, only a transpacific contact remains as a possibility. The position of the Olmecs on the Gulf of Mexico presents no difficulty, since the land connection to the Pacific is short and without natural obstacles. Olmec traces were also discovered on the Pacific coast. In fact much points to Asiatic, especially Chinese, influences. The very realistic figures represented by the Olmecs exhibit a pronounced Mongolian strain, the most pronounced in Middle and South America. The cult of the jaguar introduced by the Olmecs seems to be connected with the cult of the tiger in China. There are no tigers in America, yet the jaguar is an excellent substitute. Nor do we find anywhere else an appreciation of jade of the kind that is known in Mexico and China. The perfect working of this mineral, apparent in even the oldest figurines, also points to China. The Olmecs must have deliberately looked for jade since in their region there are no quarries where the bright green translucent stone could have caught their eye. On the contrary, jade is found there in river rock and can hardly be recognized. Only someone who knew exactly what he was looking for could find jade as it appears in this area.

Among the artistic motifs we find the two-sided representation of animals (bilateral division), man-protecting animal god-figures, helmets in the form of animals, two-headed dragons—all motifs that originated in China at the time of the

An example of the similarity in border designs from China and Central America. Ornamentation on a bronze vase from China (Chou Dynasty), top, and from a Totonac stone relief (Mexico), below.

Chou and that were preserved there for a long time. The Totonacs, who were thoroughly influenced by the Olmecs, have left ornamental border designs which an artist of the Chou could have created. Even the tree bark clothing of the Olmecs, described in later accounts by their neighbors, is probably not an independent invention. Most trees produce a raw material that is wholly unsuitable for this purpose. Only the fine fiber between the bark and the core of the trunk is suitable. Here too Chinese models might have been available. Little can be said about the calendars and the writing of the Olmecs since only remnants have been preserved. It is no longer possible to trace the Olmec origins of the much more recent Mayan and Aztec calendars.

To be sure indications of Chinese influence are by far not as numerous or clear with the Olmec as with the Chavin culture.

Nevertheless much points to the probability that their original homeland is to be sought in China. Their sudden apparition can be explained by the arrival on the coast of Chinese craft whose crews, because of their far superior stage of culture, assumed dominion over the aborigines. Whether we are warranted to include the great earth pyramid at La Venta as an example is doubtful. Scarcely to be doubted, on the other hand, is the kinship of the more recent Mexican civilizations (Maya, Aztec, Mixteca) with those of southeast Asia. There are two pyramids that can be seen in Tikal (Maya) and in Angkor (Khmer). They are almost interchangeable. In both a temple stands on the fourth and last step which can be reached by a staircase. Both staircases are decorated with serpent balustrades. The temples at the top have functionless, purely decorative semi-columns, atlantes support projections and platforms while water-lily friezes serve to structure open surfaces. The oft-cited resemblance of the Mexican pyramids to those of Khmer in Angkor (Cambodia) is problematic, since the structures at Angkor were erected around 890 A.D. while those on the other side of the Pacific are several centuries older. Nevertheless, both of these temples presumably have the same prototypes. In 1956 a step pyramid was found in Yang Tzu Shan (province of Szechuan). The quadratic structure was only 11 feet high, yet the lowest step measured the considerable length of 114 feet, the second 75 feet and the topmost 35 feet. This pyramid once must have served as a place of worship and is believed to stem from the time of the Early Chou. It was definitely built before 500 B.C. since tombs were found in it that were subsequently dated as belonging to the period of the Contending States (480-211 B.C.).

The pyramid is only the most striking symbol of the cultural relationships between Mexico and southeast Asia. The specialist in central American culture, Gordon Ekholm, has cited still others: The trefoil arch, the sanctuary within a temple, the motif of the Sacred Tree or Cross, that of a God holding a lotus flower, the particular arrangements of pillars, types of vaults, phallic symbols, diving gods, serpents deities, and bells made of wire. All these listed features are explainable if we postulate at least two transpacific impulses. An earlier one, around the time of the birth of Christ, stemmed either directly from Chinese coastal regions or from the centers of the Dongson culture, which was strongly influenced by China, and was manifested in the Olmec culture. Another very intensive impulse must have originated from Indochina. There are even clear motives for this since in 700 A.D. the missionary zeal of Indian Buddhists was especially great, and they attempted to spread the word of the Gautama throughout the whole world. That we recognize no influence of Buddhism among the traces of ancient Mexico should occasion no surprise; experience tells us that a brief missionary activity leaves no enduring influences behind. The Indian merchant sailing vessels of this time were extraordinarily seaworthy and undertook extensive voyages across the open sea up to the coasts of East Africa. It would not have been difficult for them to cross the Pacific.

The source of the northwest coastal civilization is considerably harder to locate. The tribes on the coasts of British Columbia, so different from their neighbors, first came into contact with the white man in the eighteenth century. Since they did not preserve any writings, it is often impossible to tell whether they developed a skill on their own or whether they learned it through traders, trappers, or neighbors. All we know about the iron utensils observed by Captain Cook is that apparently they did not stem from Europe. By the time ethnologists began to systematically concern themselves with these tribes, these iron instruments were no longer extant. Where they had come from and where they had gone, nobody could say. At that time they still knew copper which is used less for tools than for ornaments and the valuable "coppers," flatly hammered copper disks to which great value was ascribed. Consequently it was considered as especially prestigious to break it in the frenzy of a potlach or to hurl it into the sea. Actually it was a Stone Age culture with a grafted-on knowledge of copper. It is perhaps presumptuous to draw conclusions on the basis of this general information, yet it may be pointed out that copper always enjoyed a very high repute in China and was used there exclusively as coinage metal. The thousand-year old Chinese monetary unit "cash" was a string with copper coins.

The seaworthy double canoes with the Venetian blind-like sails also point to China, where similar seacraft were used. In addition, the tribes of the northwest coast excelled in an art already mastered by the inhabitants of China: they produced superb cloth from the inner bark of the cedar which they traded with their neighbors at a great profit. Blankets of this material mixed with the wool of mountain goats are still a coveted collector's item.

If we compare the artistic motifs, the heads and figures stacked on each other immediately stand out. The Chinese at the time of the Shang dynasty decorated bones with these figures. On the Northwest Coast they appear on the totem poles mentioned earlier. The reason that no old totem poles have been found in America or in China is simple: wood is a perishable material. There are peoples in New Zealand, Sumatra, New Guinea, and the Phillipines who even today place similarly formed totem poles next to their assembly houses. The oft-mentioned bilateral representation of animals, which experienced a remarkable artistic highpoint and indeed became a leitmotif, is known both from the Amur region and from ancient China. In the same regions artists frequently painted eyes or faces in the joints and hands of human figures—as did the Tlingit, Haida, and Chimmesyan. Still other geometric motifs (concentric, rounded-off rectangles, frog-symbols) suggest a connection to the China of the Shang dynasty.

A second tendency, however, may also be seen, which points to the Dongson culture area. To this belong the great wooden houses with totem poles (Assam, Sumatra, New Zealand), bell-shaped pestles with special handles (New Guinea, Polynesia), clubs fashioned of stone or whalebone (Maori), and shells as inlay material in wood (New Zealand, Polynesia).

These influences are extremely difficult to date. A further difficulty lies in the fact that the Northwest culture, conditioned by its isolated situation, seems to have long periods of a completely autochthonous development behind it, during which the strict caste system and the peaceful but ruinous competitions among chieftains presumably arose. We know of no other culture-area in which one overcomes a foe by destroying as much of one's own property as possible.

The few items of evidence we have speak for a transpacific contact around 1000 B.C. with China, and another one, around 500 to 800 years later, with Indochina. The idea is not so far-fetched. Vessels sailing along the coast had to pass by this region; in addition, the Japanese current originating off the Chinese coast would have propelled a vessel across the north Pacific to the northwest coast of America. Moreover, the constant westerly winds would have considerably supported this drift.

We began our excursion into American history with the question: who actually discovered America and when? In order to answer this question we considered three cultural areas that stand out from the otherwise uniform and homogeneous American historical background. The Indian cultures on the Pacific coast of British Columbia and of southern Alaska, in the Isthmus of Tehuantepec, and in the central Andes highland, had little in common with their neighbors. In addition the last two emerged quite suddenly at a fully advanced level.

Could these cultures have originated from the aboriginal population through evolution? The peoples who in several settlement waves, which began at least 60,000 years ago, streamed across the Bering Strait into the uninhabited double continent brought with them Mongoloid, Caucasoid, and Negroid racial traits, which we also find in the early civilizations. But one thing was certainly not to be found in their luggage: the trait of an advanced stage of civilization. Indeed, from 3000 B.C. on, the water of the Bering Strait constituted an insurmountable barrier for massive migrations.

From that time on the American continent seemed to be complelely isolated from the outside world, yet the Olmec and

Chavin civilizations developed in what is now Mexico and in northern Peru. How could this conglomeration of Stone Age gatherers, hunters, cultivators and basket-makers reach a cultural level within a time span of two thousand years that had required at least three to four times as long for the peoples on the Nile, on the Euphrates and on the Tigris, and on the Indus? No traces could be found which point to a continuous development of the aboriginal population. It is a characteristic of the most ancient civilizations of America (Chavin and Olmec) that they emerged without going through any period of transition whatsoever: carriers of an advanced culture must have given the impulse to this development.

Even if we set aside the problem of the Olmecs and the Chavin, there are still other indications that the American continent did not exist in such splendid isolation as was long assumed. The gourd and the coconut palm, examples of cultivated plants found in America long before Columbus, came from their original homeland in Asia but could not have crossed the waters of Pacific naturally propelled by wind and currents. Human beings must have brought them from the other shore. Doubt no longer exists on this score. And hence men landed on the coasts of America long before Columbus. Unfortunately they left no direct traces behind.

The carriers of these advanced cultures could not have come either over postulated land bridges in the Atlantic or the Pacific or over the old route of migrating peoples, the Bering Strait. They must have used the sea route. For the time of the earliest known contacts (tenth to ninth centuries B.C.) the old cultures of the Mediterranean countries have been considered as possible sources, but we have seen that transatlantic transmissions, if they took place at all, could have had only a sporadic character. Several cultural spurts are known to have occured in the central Andean region and the intensity of the influence indicates long and thorough contact, so that there remains only the hypothesis of a transpacific cultural transmission. The expanse of the Pacific that had to be overcome is partly balanced by the prevailing winds and currents which would have considerably favored a voyage from Asia to America. Direct proof of this is offered by the Asiatic ships that are repeatedly driven onto the California coast by wind and weather. The return voyage is assured by the southeast trade wind and the equatorial current. This sailing course south of the Hawaiian Islands was also used by the Spaniards more than two hundred years ago for the traffic between Mexico and the Phillipines.

The oldest traces of the Chavin culture in northern Peru indubitably point to influences from the Chinese Chou period. The congruence is often so staggering that the Chinese piece seems to have actually served as a prototype for the one found in the Andes. At the time of the oldest contact China could look back on a long history; this was the time of the transition from the Early to the Late Chou dynasty. The individual kingdoms within China achieved a large measure of independence, as did the States Wu and Yüeh which occupied the Chinese coasts between the peninsula of Kwantung and the region now known as Canton. These were to a great extent independent states, but the Chou culture flourished here also. Like most inhabitants of coastal regions the people of this area were also superb seafarers. We know that they used sailing rafts, double canoes, and presumably also plank boats. The Chinese annals provide scant information about these coastal peoples (and after all, three thousand years have since passed); they merely note that these peoples often undertook extensive military campaigns across the sea. We should not imagine that from the outset the ships of Wu and Yüeh entrusted themselves to the westerly winds for the long passage to America. They surely were the first who sailed along the coasts and are perhaps the very first who pushed forward into the region that today is occupied by the tribes of the Northwest civilization. (Only thus could we explain the few traces found there that indicate contact with the Shang and Early Chou dynasties.) From there the prevailing winds favored a voyage in a southerly direction to Central and South America.

We can be sure that any planned trips to the Americas were

preceded by involuntary landings. At that time fishermen were no doubt blown off course onto the American coast by storms, and many of them probably succeeded making the long journey home. Their accounts surely served as a spur to planned Pacific crossings. The next step then would have been to replace the coastal voyages by a shorter direct route. The time for these first contacts fits chronologically; the transmitted cultural elements came from China; the technical possibilities for this existed: seafarers of the states of Wu and Yüeh gave the outside impulse to the development of the Chavin culture. There is still another clue: judging from excavation finds in South America, the previously existing connections broke off around the fifth century B.C.; just at the time when the Wu and Yüeh kingdoms were sucked into the vortex of the struggles for hegemony in China. With that the basis of the earliest transpacific voyages was destroyed.

Around the same time, however, another, related cultural area arose in the vacuum that had been created. Bordering on Yueh in the south, the Dongson culture originated in the Gulf of Tonkin and in North Vietnam. It encompassed a series of cultural currents, among them the Caucasian elements transmitted through the Pontic migration, and radiated in all directions. We find its influences in the Malayan archipelago, in Melanesia, and even among the Maori in New Zealand. Refugees from the kingdoms of Wu and Yüeh surely brought with them not only cultural elements of the late Chou, but probably also knowledge of transpacific voyages. We again find cultural influences of the Late Chou wherever the exstence of the Dongson is demonstrated.

We do not know what motives led the Dongson people to sail to America after the fifth or fourth century B.C. but there are indications that they did do so. In the epochs following the Chavin in the central Andes region they left behind so many traces that we are forced to conclude that there were long contacts. Thus it is not surprising that the Chinese influences again become pronounced at the end of the Dongson period (around 100 A.D.). In the Middle Kingdom the Han pursued a policy of the open Chinese Wall, and an intensive cultural exchange ensued. Now the transpacific voyages from the region of Tonkin also touched regions in Ecuador and Columbia. One of the last expeditions must have landed on the Isthmus of Tehuantepec and founded the Olmec culture there. That was already at a time when the armies of the Han had conquered the core regions of the Dongson sphere. We next find the influence of the Chinese with the Olmecs. In the beginning the Dongson region was only nominally subject to China, nevertheless many Chinese settlers streamed into the country from the first century A.D. on, and the conquered regions lost their independence and became centrally ruled provinces.

The State of the Late Han was beset with many internal difficulties; the mandate of heaven was running out. Thus it is not surprising that transpacific contacts should have broken off.

In the seventh century A.D. strong impulses from Indochina and India seem to have operated on Central America once more, but thereafter knowledge of the civilization on the other side of the Pacific in Asia was completely lost. It survived only in myths and fables.

Thus it was not Columbus who was the first to steer towards America. Nevertheless, in the meanwhile, his name has become a synonym for the discoverer of America. Anyone who finds the foregoing hypotheses convincing, then, should have no qualms about asserting that "Columbus was Chinese."

APPENDIX

SIMPLIFIED PRESENTATION OF CHINA'S DYNASTIES*

HSIA KINGDOM		2000 – 1520 B.C.
SHANG KINGDOM		1520 – 1030
CHOU	Early Chou	1030 – 722
	Ch'un Ch'iu	722 – 480
	Warring States	480 – 221
FIRST UNIFICATION:		
Ch'in		221 – 207
Earlier Han		202 – 9 A.D.
Wang-Mang interregnum		9 – 23
Later Han		25 – 220
FIRST PARTITION (THREE KINGDOMS):		
Shu		221 – 264
Wei		220 – 264
Wu		222 – 280

*After J. Needham, *Science and Civilization in China,* New York, 1961.

SECOND UNIFICATION:	
Western Chin	265 – 317
Eastern Chin	317 – 420
Liu Sung	420 – 479
SECOND PARTITION (NORTHERN AND SOUTHERN DYNASTIES):	
Ch'i	479 – 502
Liang	502 – 557
Ch'en	557 – 587
Northern Wei	386 – 535
Western Wei	535 – 554
Eastern Wei	534 – 543
Northern Ch'i	550 – 577
Northern Chou	557 – 581
THIRD UNIFICATION:	
Sui	581 – 618
Thang	618 – 906
THIRD PARTITION:	
Wu Tai	907 – 960
Liao (Ch'itan Tartar)	907 – 1125
West Liao (Qara-Khitai)	1125 – 1211
Hsi Hsia (Tangut)	990 – 1227
FOURTH UNIFICATION:	
Northern Sung	960 – 1126
Southern Sung	1127 – 1234
Chin (Jurchen Tartar)	1115 – 1234
Yuan (Mongol)	1260 – 1368
Ming	1368 – 1644
Ch'ing (Manchu)	1644 – 1911
Republic	1912 – 1949
Peoples Republic	since Sept. 1949

Following page : Chart of Migrations (See page 227).

MIGRATORY WAVE	ORIGIN	RACIAL TRAITS	MAJOR EXCAVATION SITES	
I 60,000–25,000 B.C.	Northeast China	Neanderthaloid	Texas, Idaho, California, Alberta, Nevada	Stone Age hunters, very primitive stone tools. Ice-free corridor to the interior of America.
II 25,000–20,000 B.C.	Altai, Ordos	Caucasoid Mongoloid	Yukon Territory Brooks Range, Alaska	Scrapers, parers, Mousterian points. No access to the interior of America.
III 20,000–15,000 B.C.	Lake Baikal	Mongoloid Caucasian influences?	Alaska	Double-edged blades, West European elements in the working of stone, Sandia, Clovis, and Lerma periods. No access to the interior of America until 1000 B.C.
IV 6000 B.C.	Gobi, Manchuria Amur, Japan	Mongoloid	Alaska, British Columbia Alberta	Microblade culture; America ice-free as far as the high mountain regions and the North. Bering Strait begins to break up.
V 5000–4000 B.C.	Siberian Tundra	Mongoloid	West Coast of Alaska	Arctic microlith culture, later spreading to Greenland. Stone lamps harpoons. Bering Strait divides the two continents.
VI 4000 B.C.	North Pacific coast of Asia	Mongoloid	Northern shore of the American continent	Eskimos, Chuckchees. Kayaks.
VII 3000 B.C.	Siberia	Mongoloid Caucasoid	The entire North American continent, Algonquin Indians	Algonquins: introduction of pottery making from Asia. Does not appear in interior of the country until 2000 B.C.
VIII 1500 B.C.	Asia	Mongoloid	Athabaska Indians, Northern Canada as far as the prairies	Later penetration as far as Mexico and Peru.

Bibliography

Beazley, K. *Dawn of Modern Geography.* New York, 1897.

Binns, N.E. *An Introduction to Historical Bibliography.* London, 1962.

Bowker, H.F. "A Note of the Chin Dynasty." *The Numismatist,* 63 (1950), 313.

Brown, L.A. *The Story of Maps.* London, 1951.

Cable, M., and F. French. *The Gobi Desert.* London, 1942.

Carter, G.F. "Plants Across the Pacific." *American Antiquity,* 18 (1953), 62.

Carter, T.F. *The Invention of Printing in China and its Spread Westward.* New York, 1955.

Chang, Kwang-chih. *The Archaeology of Ancient China.* London, 1963.

Chidsey, D.B. *Goodbye to Gunpowder.* New York, 1963.

Cipolla, C.M. *Guns and Sails in the Early Phases of European Expansion, 1400-1700.* London, 1965.

Covarrubias, M. *The Eagle, the Jaguar and the Serpent.* New York, 1954.

Dawson, R.S. *The Chinese Chameleon.* New York, 1967.

———. *The Legacy of China.* Oxford, 1964.

Duyvendak, J.J.L. *China's Discovery of Africa.* London, 1949.

Ekholm, G.F. "A Possible Focus of Asiatic Influence in the Late Classical Cultures of Mesoamerica." *American Antiquity,* 18 (1953), 72.

———. "Is American Culture Asiatic?" *Natural History Magazine,* 59 (1950), 344.

Gibbs-Smith, C.H. *The Aeroplane.* London, 1960.

Gladwin, H.S. *Men Out of Asia.* New York, 1947.

Goodrich, I.C. *Short History of the Chinese People.* New York, 1943.
Isis, 36 (1946), 114.

———, and Feng Chia-Sheng. "The Early Development of Firearms in China."

Hart, C. *Kites.* London, 1967.

Hedin, S. *The Wandering Lake.* New York, 1940.

Heine-Geldern, R. "Significant Parallels in the Symbolic Arts of Southern Asia and Middle America." *Sel. Papers 29th Cong. of Amer.* (1951).

Herrmann, A. *An Historical Atlas of China.* Chicago, 1966.

Hobson, E.W. *Squaring the Circle.* Cambridge, 1913.

Honore, P. *In Quest of the White God.* New York, 1964.

Hudson, G.F. *Europe and China.* London, 1913.

Imamura, Akitune. "Tyoko and his Seismoscope." *Jap. J. of Astr. and Geophys.,* 16 (1939), 37.

Ling, Wang. "On The Invention and Use of Gunpowder and Firearms in China." *Isis,* 37 (1949), 160.

Lum, P. *The Purple Barrier.* London, 1960.

Miyasita, Saburo. "A Link in the Westward Transmission of Chinese Anatomy in the Later Middle Ages." *Isis,* 58 (1967), 486.

BIBLIOGRAPHY

Needham, J. *Within the Four Seas.* London, 1969.

———. "The Roles of Europe and China in the Evolution of Oecumenical Science." *J. Asian Hist.,* 1 (1967), 3.

———. *The Development of Iron and Steel Technology in China.* Cambridge, 1964.

———. *Science and Civilization in China.* New York, 1954.

Parks, G.B., ed. *The Travels of Marco Polo, based on the Standard English Version.* New York, 1930.

Partington, J.R. *A History of Greek Fire and Gunpowder.* Cambridge, 1960.

Quiggin, A.H. *A Survey of Primitive Money.* London, 1963.

Reinfeld, F. *The Story of Paper Money.* New York, 1958.

Rostovtzeff, M. *Caravan Cities.* Oxford, 1932.

Schroeder, D.L., and K.C. Ruhl. "Metallurgical Characteristics of North American Prehistoric Copper Work." *American Antiquity,* 33 (1968), 162.

Silverberg, R. *The Long Rampart.* Philadelphia, 1965.

Stalker, A. "Geology and Age of the Early Man Site at Taber, Alberta." *American Antiquity,* 34 (1969), 425.

Stein, M.A. *On Ancient Central Asian Tracks.* London, 1933.

Wallacker, B.E. "The Siege of Yupi, A.D. 546." *J. Asian Studies,* 27 (1969), 789.

DATE DUE

DEC 10 '75

GAYLORD PRINTED IN U.S.A.